The Book of OSM
© 2015 Steve Coast
Edited by Barbara Poore

With special thanks to Kickstarter backers, in order of backing date:

Story Editors:

James Fee, aojajena, Ed Freyfogle, Jonathan Hipkiss, Ryan Peterson, Ed Parsons, Bojan Sobocan, Rob Dunfey, Rob Fairhead, Glen Braun, Keisuke Nakao, Chris Bol, David May, Tenebrae, Henk Hoff, Philipp Schneider, Çağatay Çelebioglu, Alexandre Leroux, rweait, Jaroslav Bengl, Roy Mitsuoka, Alan Glennon, Milo van der Linden, Brian DeRocher, ikiya, jimmyrocks, Steven Ramage, Nick Hocking & Neil Cawse.

Editors:

Christopher Jones, Joshua Kraayenbrink, Tom Elsner, mikel, Christopher House, Adam J. Gardner, Benjamin BERNARD, Klas Karlsson, suad, Philippe Ameline, Peter Papp, Ilja Zverev, Ed Summers, Andrew Owen, Jaakko Helleranta, Morten Bek Ditlevsen, Levente Juhász, Russell Nelson, John Lauck, Alexey, Dongpo, arnaud, Willy Bakker, Mahesh Prakriya, Morgan, Paul Mach, Laurence Penney, Finn Newick, Robert Barr, Erik Escoffier, Matteo Starri, Hippolite Bureau, Johannes Kebeck, Stewart C. Russell, Maxim Dubinin, James Crone, Randy Meech, David Nesbitt, Taichi FURUHASHI, Toshikazu Seto, Jóhannes Birgir Jensson, Blu, Ariann Nassel, HK Hamai, Stefano Lippolis, Wolfgang Sprick, Jérémie Salvucci, Marc Levy, Daryn Nakhuda, Felipe Knorr Kuhn, David Lang, Darrell Fuhriman, Bas de Bakker, Peter Hoeflehner, Takuya Shimada, Darren Cooper, Seth Fitzsimmons, Victor Ferreira, Peter Batty, David Sonnen, Sascha Brawer, Ben Discoe, Sam Iacullo, Victor Chen, Ian Dees, Luis Fernando Chavier, Richard M, Markus Mayr, David Curran, Kristen K, Peter Aronson, Ben Gimpert, "Shahram Rezaei, Ben Tuttle, Jarno Peschier, Domenico Santarsiero, Matt Oakes, Edward Mac Gillavry, Sam Franklin, Christopher Lorenz, Thomas Bähler, A. Lee, Tomas, Alan Meban, Paul Robertson, Alasdair Nicol, Teri Centner, Bill Bonwitt, Ron H, Kirk Bradley, TC Haddad, Richard Woodruff, Daniel Ceperley, Ethan Trooskin-Zoller, Kai-Florian, Hellebaard, Gregory Marler, Jeroen Muris, Robert Surtees, Arnaud Levy, Alex Hamling, Stefan Langer, Naxo, Robert Stack, shaun, Charlie Langridge, Colleen Kaman, Nobiloo, Luke Philips, Darren Wiebe, Alphonso Dandridge, Eduardo Larrosa, vmeurisse, Xavier Barnada Rius, David Colhoun, Elijah Meeks, Kristofor Carle & Brendan Devenney.

Associate Editors:

Jan Ainali, Beau Gunderson, Robert Nordan, Thomas Ungricht, Peter ter Haar, Frank Wales, Orion, 飯田哲, Phillip Stewart, steppkestift, Heather Leson, Tim Balzer, Andrea Borruso, Nuno José Almeida, Marc Metten, Kenneth Hurst, Mark Murphy, Jez Higgins, moobsen, Veli-Heikki Vesanto, Chris MacWhorter, Jesse Rouse, Ricky Curioso, Zdenek Henek, Jonathan Gross, Karl Ove Hufthammer, Nikket Pokharrna, Christian Marc Schmidt, Dan Karran, Jens Nitzschke, Satoshi Kinoshita, Tom Curtis, Igor Tihonov, Luc Vaillancourt, Mark Parthum, Tom Landry, Matt Sweeney, Ron Lansverk, Carl D. Patterson, Karl-Petter Åkesson, Vincèn Pujol, Al Pascual, paulswartz, Zian Choy, GPSPython, Robin Rattay, Roberta, Will Knight, Jaime Crespo, Manfred Fuchs, Patrick Singer, Thejesh GN, 2odrigo Perez [Mx], James, Kevin Bullock & Isaac Best.

Assistant Editors:

Britta Gustafson & Cedric Howe

Table of Contents

Introduction 2
Andy Allan 3
Tom Carden 16
Kate Chapman 40
Steve Coast 63
Martijn van Exel 99
Richard Fairhurst 116
Ben Gimpert 133
Sean Gorman 146
Muki Haklay 166
Henk Hoff 183
Pavel Machalek 196
Mikel Maron 200
Randy Meech 222
Ed Parsons 239
Andy Robinson 257
Grant Slater 274

Introduction

This is a book of interviews with people who were around when OpenStreetMap was started. It's designed to provide a variety of viewpoints on something that happened over a period, roughly, from 2004 to 2014.

It would take too long to interview everyone with partial responsibility and thus this is a partial list of those important to the project. What you find here are historical snippets, anecdotes and points of view on an audacious project: Map the world with volunteers and give the data away for free. This is something that was incomprehensible even a few years before the project was started.

The book is designed to be slightly raw. It could have been edited to highlight my personal genius and so on, but instead I hope it gives a feeling of fallibility and the ongoing experimental nature of the project. Not a vision from heaven faithfully engineered but a messy project in every sense.

Each interview also includes a short biography of the interviewee and a quote about their favorite map. It's deeply ironic, but securing copyright for each of these is beyond the scope of my sanity and thus links are included instead.

All the interviews except my own were conducted by me. Mine was conducted by Adam Fisher, noted journalist from Wired and elsewhere. The project was managed by Ben Wroe and edited by Barbara Poore. Many thanks to each of them.

All mistakes are mine!

Steve Coast
Founder, OpenStreetMap
Denver, Colorado, October 2015

Andy Allan

Bio

Andy Allan is a digital cartographer based in London, UK. He joined OpenStreetMap in 2006 and has been a mapper, developer, cartographer and operations volunteer ever since. His business thunderforest.com creates high-quality customised maps for developers and businesses around the world, based on OpenStreetMap data.

Favorite Map?

"My favourite map is the OpenCycleMap layer in Putney, London. It's where I first started mapping with OpenStreetMap, and was where I initially experimented with creating my own custom maps showing the cycling information that is otherwise hidden. OpenCycleMap rapidly grew from just showing Putney to cover the whole world, and is now used by tens of thousand of people every day."

http://www.openstreetmap.org/#map=16/51.4619/-0.2170&layers=C

Interview

Who is Andy Allan and what were you doing before you found OpenStreetMap?

Well, this is back in 2007. I was working for a university as an IT support engineer but I've always had an interest in maps and mapping. The thing that got me into OpenStreetMap was, one summer I was walking the Pennine Way which is a long distance foot path in the U.K. It's about 270 miles long. My friend had a Garmin GPS with him and we could see the path that we'd been on because it left bread crumbs showing you where you've been but we didn't have any maps for it because they were so expensive. So I fancied buying a GPS but I needed a good excuse and I'm also interested in open source things. I was wondering if there were any open source maps. That's where I first came across OpenStreetMap and that was both a project where I contribute into and an excuse to go out and splash £100 on a GPS.

How did you first find OSM?

I found OSM through Google but I had seen it mentioned a couple of times previously on some open source software development blogs. So yeah, that was how I first found it. And when I started, there was nothing there. There were 40 or 50 roads in London and nothing anywhere near where I lived. And so, I took my GPS and walked around the block, came back in, and tried getting the software working to see if I could add those streets around my local block onto OpenStreetMap. And it all began from there.

How much of a success was it?

Oh, it was a success. It was possible. I quickly got addicted to it and within the first couple of weeks, I'd mapped most of the streets around where I lived, on foot, and realized that I needed to get a bike mount for my GPS so I could map more of the local area. This was in Putney in Southwest London. At the time, I was sharing a flat with a couple of friends and one of them, Dave Stubbs, also got interested and bought his own GPS and came out and helped me map.

Do you want to talk about the process of mapping and what the mechanics of it were?

Yeah, the mechanics of it back then when I started was that we had nothing to help us other than GPS. So we would set the GPS to log where we were every second. And so, it would record the position you were in. And as you walk up and down the streets, it would leave a trail of dots behind you--on the GPS. We'd also note down things that we saw when were out and about, primarily street signs, the names of roads in the local area. We'd then come back to our flat and use the editor.

There were two editors. There was an applet—a Java applet which I think you had a hand in. It showed very low-resolution satellite imagery, not anything useful for mapping. But then it could also show the GPS tracks so that you could see where you've been and start drawing the lines to represent the streets over the top of it. There was also another editor called JOSM which is still running, but that would be the essential piece of software to install on the laptop. Looking back now, from seven or eight years in the future, it seems very primitive but the principles are almost exactly the same.

Where did you go from there after mapping your local area?

Mapping more and more of Southwest London. We were at the point where we had mapped all the roads within a 30-minute bike ride radius of our flat. So it was becoming increasingly time consuming to map, but our interest was sparked because we had to map out every street, there was nothing left to start with, we were mapping every street.

And as we were going around, we were finding that some streets had sign posts on them for bicycle routes. And at first, they made no sense at all because they were just pointing down back streets and then there'd be no sign further down the road. But because we were systematic, we were going up and down every road we would find that some of these kind of made sense. If you checked every road, you'd find the next sign post and realize, "Ah, okay, so there's one missing on this corner but this is where the route goes." And so, we were adding this information about the cycle routes into OpenStreetMap but had no way of visualizing what was going on. I started working on a custom map in order to show the cycling infrastructure, to show where these cycling routes were, going around the streets and around the paths in the local area. And that distracted me very quickly from completing the rest of London but thankfully, by then, there was plenty of other people involved to do more mapping.

Do you have any other early memories of the project?

I'd been reading the mailing lists for a while and I think I had signed up to the mailing list. I'm not sure if I posted but you, Steve, were organizing a series of talks called Ask Later and I remember going along to one of them. It was the first time that I met anybody from OpenStreetMap so I just went along, sat in the auditorium, listened to the talks, and didn't ask any questions. Went to the pub afterwards and immediately was offered a beer by Etienne.

Etienne was one of the main guys in OpenStreetMap. Back at that time, he was writing editors and writing lots of software. I was becoming rapidly star struck because there was Etienne, on one side, offering me a beer. Steve Coast was there. There was Richard Fairhurst. Basically, half the names of the OpenStreetMap project were all in the pub and all completely welcoming and happy to talk about anything that I wanted to talk about. We even had Richard who was showing off the very first preview of his new editor called Potlatch. That was definitely eye catching because it was so much easier to use than the Java applet that was there. But after that meeting and the meetup afterwards, the biggest thing was just how friendly and how welcoming everybody was. All of a sudden I didn't feel I was just some mapper and that gave me a lot of encouragement to continue.

I remember those. I think they were held at Westminster University or a lot of them were.

I think I only went to one of them but Etienne was talking about health records and the new Open Health Record thing. There was a talk on building a statue to someone and I can't remember who it was – some Linux guy. I think the narrative was on the wings of when you walk around Westminster you see all these statues to show soldiers and generals and things like these. Can we not have some statues for computer people? I can't remember how many I went to. I think I only went to one of them.

What happened after that? What was your involvement in the project?

The map that I created rapidly became what I was doing in OpenStreetMap. It was originally called the "Cycle Map" but the founder of OpenStreetMap suggested to me one day to register opencyclemap.org and that's caused years of confusion ever since.

Why is that?

Well, the confusion with OpenCycleMap is that when people see the maps I've created, they think that's a parallel project or a duplicate of the OpenStreetMap project when in fact all the data that gets shown on the map is part of OpenStreetMap. It all comes from the same OpenStreetMap database. So I have a lot of inquiries saying things like, "I've downloaded the OpenStreetMap data. Can I download the OpenCycleMap data because I want to do my own thing with bike routes?" And I have to explain that they're not parallel projects. It's just a custom rendering of the same OpenStreetMap database.

So anyway, I was doing this cycle map and then it became OpenCycleMap. At that point, I was still working for the university. But this leads into the forming of CloudMade. You and Nick had got funding for CloudMade and approached me at one of the Hack Weekends to ask if I wanted to come and join and bring my cartography expertise into the company and I said, "Yes." And so, that took OpenStreetMap from what started as a hobby and then became a very time-consuming hobby, and it became both a hobby and a full-time job.

Before jumping into that, do you want to talk about Hack Weekend, what happened, and any memories?

I remember a couple of Hack Weekends before joining CloudMade. Held at offices in Putney which was a nice coincidence because it meant that it was only 15 minutes walk from where I lived. On one of the Hack Weekends, and I think it was before CloudMade started, we were developing a new version of the API that included relations and that meant that we were having to change the Rails Port software—the Ruby on Rails codes that runs both the OpenStreetMap website and this map editing API. And so, changing that software, changing the database layer and changing of editing software, all at the same time. But we had people who were working on all the different bits of software in the room. We split into multiple rooms to work on different tasks but that's far back as I can remember and I can only vaguely remember that one.

I do remember of our mapping parties that I went to, there was one held in the MultiMap Office in Central London where we had—well, it must've been about 20 odd people there. I missed the first day because I was on a holiday and came back straight from the airport to the mapping party and went out to help map the West End of London. And then, in this stage, we still didn't have all the streets in Central London even around Mayfair and Oxford Circus. There were only the main roads. We still had to get all the back streets.

I also remember another mapping party around the same era, I think just six months later, at a sustainable living complex called BedZED in Beddington in

South London. And there, we had a mixture of both streets to map, and also this vast wasteground where we were wandering around trying to figure out where the paths went and avoid all the locals who were on motocross bikes and surprised to see somebody walking through that area. But I remember it had been a tradition to go to the pub afterwards. There were only two other people other than me who felt like going to the pub. That was my flat mate, Dave, whom I had done lots of mapping with and you. And so, the three of us went to a fairly run-down pub nearby and sat there for several hours drinking heavily. And that was another good day. And, again, just nice and welcoming that the founder of OpenStreetMap was happy to sit in the pub with two relative nobodies. Yeah, I remember that.

Then there's CloudMade, to discuss. I was there for 18 months. That was a very intense time. I don't really know where to start. I agreed to join CloudMade. I knew that quite a lot of the other people who were heavily involved in OpenStreetMap and based in London, had also been asked. And I thought this was a fantastic opportunity to do what I'd been doing, but being paid to do it. And so I was slightly surprised that some other people didn't want to join CloudMade but, yeah, I was excited at the opportunity.

When I started, I was the first employee of the company to start after the two founders, Steve and Nick. I was the first employee to be able to work there. Other people were serving notice and other people had been hired afterwards. So I got to see the transformation from a spacious two-person office into a slightly cramped six or seven-person office a couple of months later.

The start of it, everything happened in the first week. I was sent to buy desk fans and printing supplies from the Ryman across the bridge in Putney. And back and forth, buying post-it notes because by the end of that week, there was a CEO and a CTO and advisors sort of flown in and were in the meeting room and we needed to buy things like whiteboard markers and things like that.

And then a week later, I was on a plane out to Kiev in Ukraine to brief the tech team out there. Basically, just to take everything that I'd learned about OpenStreetMap and to try and share it as quickly as possible with a team of—I think it was about 25 engineers over there who had known about OpenStreetMap for about a week and were desperately needing information on how everything worked in order to get going. And that was just day after day of sitting in a meeting room, talking constantly with everybody listening to what I had to say, and asking questions. So the first two weeks were completely surreal—a brief moment of having no idea on where to start and then being overwhelmed with this immense amount of things I needed to do.

Well, you seemed to have scaled well.

It was strange though because when I started at CloudMade, my job title was Technical Lead. I didn't, at first, realize that there was a CTO who had been appointed as well. That was interesting mainly because the CTO had worked closely with the Kiev engineering team for years. And so, he knew everybody there, which was a major advantage, but we had so many people in Kiev that I didn't know what their skills were and I didn't know what their plans were and what they were working on. And so things moved about quickly. It was hard to keep up with all the tasks that needed doing and who was doing them and where the lines of communication went. So it was very intense trying to get the ball rolling.

And it was a huge culture shock as well, for me. It's not a role I'd ever worked in before. I've never had that many staff or that ability to change what was going on. And I think I ended up more as a knowledge bank like "Can you do this with OpenStreetMap?" And so people were using me for that knowledge.

A lot of the decisions got made elsewhere and that became more and more apparent as the months went on. In CloudMade, the decisions were all being made by the CTO, mostly being communicated directly to the engineers and what had initially started as a major position within the company become more and more sidelined as time went on. And I think a lot of that is just due to inexperience. It wasn't something that I had done before. It would be very interesting to do the same thing again.

What happened from there?

Well, CloudMade grew quickly. We had lots of interest. There were lots of people wanting to use OpenStreetMap commercially. And of course, we were driving lots of interest as part of being a startup.

We were building a map service to display cartographic maps and that's the one thing that I had the most input into. It was the one thing that I had most experience with. We had a geocoder. We had everything.

And we had various launch parties. I remember doing a launch party in London but I can't remember the name of the place it was in - some Gothic building off of Holborn. And yeah, doing all the arrangements for that, getting screens into the right place, finding that projectors were not happy with our headline map style that Stamen—a design consultancy—had made for us. Basically, everything was washed out and that's like an hour or so before launch was due to start. These launch events, both in London and in San Francisco, were always very high stress affairs. And I've never entirely understood why because looking back, even though a few months later, like it just seemed a lot of hassle for not much reason.

I flew around the world working for CloudMade. I was attending conferences in South Africa and Norway, going to our offices in California, going to the Consumer Electronics Show (CES) in Las Vegas and to Kiev. This was quite a step change from everything that I've done in OpenStreetMap before which was more based around pubs and people rather than traveling overseas and things like that. So it was definitely the high life working for a startup.

Andy Allan

Any memories of management exercises that you want to share?

The management exercises were interesting. We had various off-sites where we would go and make sure that we were planning for what the company would do, what products were best to work on, and how we would earn money in the future if and when we ever decided to earn money. We would also do things like–they seemed ridiculous but wasn't that bad. We'd take piles of magazines and cut out pictures from the magazines that would portray the lifestyle that you want to live and make collages from it. In fact I still have one of mine just here [shows it to camera]

We'd make these collages of what our lifestyle was. None of us had been told what the next step was, but basically after you'd cut out the things that you thought gave you the right kind of lifestyle you had to put a capital price on each of the pictures that you had and then sum up how much money you would need in order to live that kind of lifestyle. And I remember sitting in the hotel, somewhere north of London, making this with everybody else in the management team and the sum of everything that I wanted was £75,000 plus also a house and that was it. That was the capital expenditure. So given U.K. prices, that's probably about £375,000. And the next nearest in the management team was over £10 million because everybody else has got private jets and helicopters and enormous palaces in California and things like that. So that made me feel like quite an outsider because I had quite a lot more modest goals. But where this fit into the company planning was that we then know how to take the capital amount and figure out how much CloudMade had to sell for in order that we would get that much money from our stock options. And so, that meant that CloudMade needed to be quite a valuable company if everybody was getting out their $100 million or $300 million of share value from it.

My memory of one of those sessions is that I started where I'm cutting out a private jet. And so, being the guy that I am, I took a look at that and decided to cut out pictures of Boeing 757's as the next logical step.

Meanwhile, I have things like a snowboard on mine because that's as far as my ambition went but considering their desired goals, I think I probably got the biggest fraction of CloudMade remuneration out of everybody else in the room, so maybe I won in the end.

What I think is a more interesting topic that came out of those meetings was whenever the discussion turned towards how CloudMade would interact with the rest of OpenStreetMap. And that's a very interesting area because, being a commercial company, we had to make money - and make a lot of money in order to be successful and pay back the investment. And that often comes into conflict with the wider OpenStreetMap community, which is almost exclusively done for non-financial reasons. And so, we had to try and figure out how those co-existed without exploiting OpenStreetMap, without dominating it, and without raising the hackles of everybody who's involved in OpenStreetMap.

And because of my background of being in an existing OpenStreetMap volunteer, and having a bit of a different viewpoint from the rest of the management team in CloudMade, that could often lead to heated discussions where it was very much me on one side of the debate and everybody else on the other side of the debate. And I remember that from one meeting in Las Vegas where we had new members on staff who just joined the management team. One of them didn't even last a week and he seemed, to me like the best hire that had been made for years, and that was again underlining the fact that we were never quite pointing in the same direction, between me and other people in CloudMade.

My favorite bits of CloudMade were the London Engineering Team, by far. We ended up with three other engineers and me—that was Matt Amos, Shaun MacDonald and Harry Wood joined us later.

We did great things in London, both for CloudMade, in terms of software that we developed, and in outreach, in being a face of CloudMade amongst the development community, and also directly for OpenStreetMap because we had the great position of being able to work full-time on really hard things for OpenStreetMap that otherwise weren't getting done. Things like organizing Hack Weekends were also easy because we could use our offices. We could get CloudMade to pay for all the pizzas. We could get people to fly in. That was great.

Big things like changing the API version. Now, I'm trying to get the technical details right. For example, we had big changes scheduled for the API that involved database downtime. Like for four days, OpenStreetMap was just switched off but because we worked for CloudMade we had the time to practice, to do trial runs of loading the database in its new setup onto hardware that was sitting in the CloudMade office and we could put Matt working on that for weeks at a time. And Shaun doing other bits of coding directly for OpenStreetMap. He'd spend about half his time just working on things we thought were good without any overview as to whether it was good for CloudMade or not. Those were great times and we did a lot of work. We did a lot of work for the United States as well. Back then, there weren't many OpenStreetMap volunteer mappers in the United States and there was a big market opportunity for CloudMade so we did use that combination to do things that would help OpenStreetMap.

CloudMade had another program of mapping ambassadors which I was never directly involved in. But the London Engineers could do other engineering related things. We were using a new routing engine and realized very quickly that you couldn't route across the United States. You couldn't get from San Francisco to New York using OpenStreetMap data because there were so many gaps in the road network. And so over a period of a couple of months and kind of below the radar, without telling any of the managers, Matt and I had worked our way across connecting together all the bits of broken road and fixing various bits of interstate in order that the U.S. routing could be done. It was great when we managed to get a connection all the way from one side of the country to the other.

Then we created a piece of software that would take the 250 largest cities in the United States and check whether you could get from one to the other on OpenStreetMap data, published a big—like a grid of which connections were possible. The kind of thing that you used to find printed in the back of a paper atlas that shows you the distances between different cities. While we were doing this, based on OpenStreetMap data and we made it available for the community to jump in.

And after us working quietly away for months, it was amazing how quickly we could get other people to jump in. And we would then think up new challenges for them. First, there was connecting all of the cities together. That was something we've managed in about three weeks. And then we started measuring the distances between them to see which ones weren't connected properly where you had to go on a long detour on OpenStreetMap data. That was good fun. That was us using engineering ability within CloudMade to make big changes to OpenStreetMap.

Do you think that a lot of things that happened at CloudMade have been repeated multiple times over the last few years?

Yes. There are lots of companies that have come along with ideas on how to make money from map-based things using OpenStreetMap and a lot of them have made the same mistakes and the same decisions that were made in CloudMade. The one thing that hasn't been repeated is the direct connection between core OpenStreetMap developers being hired and working for the companies in the same way that we had four core OpenStreetMap developers who were working on the same team and all with a quite a lot of flexibility to spend lots of man-hours working on the core software. Nowadays, it's different. It's much bigger. There's half a dozen to a dozen companies who have engineers working on OpenStreetMap and various pieces of software in the OpenStreetMap ecosystem but none of them are really paying the same focus on the core pieces of software, things like the mapping API, the database structures and things like that that we used to do in CloudMade.

The thing is that I worry about it much less now because OpenStreetMap itself is so much bigger and has so many more people involved in it that it's not as vulnerable as it used to be. One of the things when CloudMade started up was the man-hours that we had available were much bigger than the man-hours of the rest of the OpenStreetMap community when it came to technical things. And there's always the opportunity to do serious damage to the OpenStreetMap Community by doing the wrong thing.

CloudMade was also the first, so there was a lot of uncertainty about things. In CloudMade, we were creating our own OpenStreetMap editor as an alternative to Potlatch—the original Potlatch. And it wasn't clear to the community what would happen if a commercial company was developing an editor for OpenStreetMap. Would this lead to some kind of bait-and-switch? Would there be a community capture by a commercial organization? And so there's lots of discussions and heartache that went on back then.

But now, we don't have these worries anymore. We've seen how it plays out and it works for OpenStreetMap. It's not an existential threat to OpenStreetMap so we can have other companies like Mapbox creating editors without all the worry that went along when we were doing it in CloudMade.

Good, so what happened next?

Well, CloudMade had big visions on what it was going to do. But being a venture capital-funded company, there's always a strange balance to be had between whether you want to fund the company through investment or whether you want to fund it through revenue. And one of the big downsides of funding it through revenue is that you have actual numbers to plug into your spreadsheets and that might not be the numbers that you want in your spreadsheet when you're showing upper management.

So instead, what we normally did in CloudMade was to give everything away for free and then we could use whatever numbers in the spreadsheets seemed appropriate. But giving everything away for free meant that we were always at the mercy of the next funding round—when it would happen and how big it would be? And one of those funding rounds fell through and CloudMade didn't have enough money to continue paying all the engineers. And so, the London Engineering Staff were all made redundant. And so we had to pack up and move on.

But I knew that I wasn't finished with OpenStreetMap. I knew that there had been so many things that were frustrating me working for CloudMade where I believed I had a better idea than what would come out of CloudMade. And so, rather than going and working for somebody else, I decided to work for myself as a freelancer whilst I was building my own company, building my own systems and ideas of how I would make services based on OpenStreetMap.

So even before I left CloudMade, the OpenCycleMap layer had become extremely popular. There were dozens of iPhone apps, websites, and Android apps by then all using OpenCycleMap as map layers within their applications and sites. I wanted to develop that further. A lot of people thought that my main interest was in cycling maps but as that was only just one of my ideas. So I needed one to push on and create some new map layers.

The second specialist map layer that I made is the public transport map that you see on OpenStreetMap. It was using all the same software and skills but creating new map layers. I've gone on since then to create more specialist map layers like outdoors-focused maps and landscape-focused maps.

I basically would build a platform for hosting custom map layers based on OpenStreetMap. I started charging heavy users of OpenCycleMap whilst I was freelancing. And as the uses have caught on and these maps have become more widely used and more popular, that's aided my business - charging the application developers for using the maps. It has grown and grown until the point where I'm not doing any freelancing anymore. I'm just working on these services.

That's very cool.

It's interesting how different the world is now, from when I left CloudMade, because there are so many other companies who have come along since and enjoying the OpenStreetMap sphere, from Mapbox to Skobbler who are now a part of Scout. Everybody coming along with their own ideas of how to create services worth paying for based on OpenStreetMap data. And I think the most successful ones have always taken a slightly different route from CloudMade. Even back when I worked for CloudMade, I knew that these were things worth doing and that they worth paying for. And the successful companies are people who have stood up and charged money for their services. And there's plenty of people doing that now.

Money, what a controversial idea.

It's controversial in some areas like, you know, it's controversial in venture capital-backed startups but even with no investment you can earn a living. What I've done over the last few years is build a company where I can charge for the services and it works as a business.

It might be interesting for readers to try and understand that there are these different outputs to OSM use by different types of people and that the data is richer than what you just see on OSM.org.

This is the core of what I do with my cartography. It's to take the main OpenStreetMap database and pull out the information that's buried in there, that's useful for specific purposes. So OpenCycleMap was the first specialist map based on OpenStreetMap data. And that was finding all the data that cyclists are interested in, whether it's bike routes or bike parking or bike shops and showing them on the map where they otherwise wouldn't be shown and then customizing the look of the rest of the map, toning down the major roads so that they're not the most eye-catching features, adding contours and hill shading and things that are more important to cyclists than they are to regular users.

That's the theme that has gone into all the maps that I've made. OpenStreetMap is so much more rich in information than what you see on the standard map layer on openstreetmap.org. And it's my belief that it's impossible to create one single map that's useful for all purposes so therefore you need specialist maps. So whether that's public transport information, or hiking routes, or skiing routes or the names of mountains, the heights of mountains, all of these little bits of details can be shown on different maps.

So what's your involvement in OSM today? I mean, obviously, you've talked a bit about your company but what else?

Well, I do two main tasks now in OpenStreetMap. The first one is I'm the current maintainer of the main map style with openstreetmap.org. And this is a long-running project that predates my cartography involvement so it's got quite a lot of heritage. About two years ago, it had fallen into lack of development. There wasn't much going on. Meanwhile, Mapbox had developed a piece of software called TileMill which allows you to more easily edit the map "style sheets" - the technical code that controls what features do you see in what color of things that appear on the map. I had the experience of converting maps to use this piece of software that I'd been using in my own business. And so I thought it would be a good way to give back to the OpenStreetMap Community to do the conversion there.

There'd been a mailing list conversation that nothing was happening and somebody should really step up and do something, however we're going to do it. And instead of getting involved in yet another mailing list conversation, I just did the conversion and then presented it as a fait accompli. And so, that sparked the renewed interest in the map style and I volunteered to continue maintaining the map style and to try and give it some cartographic direction. But perhaps the hardest thing I've ever done in OpenStreetMap is to try and balance out the competing desires of almost everybody I know in OpenStreetMap who wants their own pet feature to appear on the main map style.

But the project's matured quite nicely. Most importantly because now we have a team of maintainers—three other guys who are all spending huge amounts of time working on it. It's a nice healthy project with lots of discussions and lots of pull requests and things. And a much better shape that when I set out to do it. So even if I don't do much more with that style, I'm just very happy the way that it panned out and taken an important piece of OpenStreetMap that was becoming neglected and then changing it back into a healthy project again.

Cool, so what should we have talked about that we haven't talked about?

Well the one more other thing, the second thing that I do is my main involvement in OpenStreetMap is on the Operations Working Group of the OpenStreetMap Foundation. I've been doing this for a few years. And again, becoming increasingly pleased with the way it pans out.

The Operations Working Group are the group of volunteers who look after running all the servers for OpenStreetMap. I think it's about 40 servers now, in various locations around the world, mainly in two universities in London and also in various data centers elsewhere. We've got a team of around six people who look after the servers, do the software configuration, and decide how much money we need to spend on servers in order to keep the whole OpenStreetMap show on the road—because without these servers running, there's no website, there's no database. It all stops working.

Recently, we changed all of our servers over to Chef for configuration management which allows us to write down exactly what software is installed on each of the servers. It's interesting technically but the more important thing is that we now publish these configuration management control files and anybody in the OpenStreetMap community who has the interest and technical know-how can now make proposed changes to the software that's running on the servers. So we're opening up. There's no longer just people who are in London and who can get into the data center to push things around that are involved, there's people from all around the world.

And my main focus on the Operations Working Group is on how do we get more people involved in the kind of core part of OpenStreetMap and reporting back. I don't make changes myself. I don't buy the hardware or shift things around. I'm looking more at how can we get more people involved.

Tom Carden

Bio

Tom Carden is a British software developer and designer based in San Francisco where he leads Square's Data Visualization team. Carden creates dynamic and interactive graphics, maps, and data visualizations for the web and mobile devices. As a co-founder at Bloom he built Planetary, a visual music player for iPad that became the first digital acquisition at the Smithsonian's Cooper Hewitt Design Museum. At Stamen Design he created innovative software for clients like Microsoft, MSNBC, Trulia, and the London Organizing Committee for the Olympic Games. Previously, he wrote passenger flow simulation software for YRM, a London-based architectural studio. He was an active and early participant in the Processing and OpenStreetMap communities and has spoken at conferences including South by Southwest, UX Week, ETech, and Where 2.0. He holds a B.Sc. in Artificial Intelligence with Mathematics from the University of Leeds and an M.Sc. in Virtual Environments, Imaging and Visualisation from University College London.

Favorite Map?

"Lake Merritt in Oakland, since that was where all Stamen's maps started"

Interview

Why don't we start off with who was Tom Carden before you found OpenStreetMap and what were you doing?

I moved to London in 2002 and found OpenStreetMap around 2004 or 2005. I was working for an architecture studio in London. I was writing simulation software for airport designers to understand pedestrian movement in airports. And so, I was an AI geek. I was a graphics geek. And I was learning essentially how to sell engineering to designers and how to work simulation and software into a design process and actually have it be a meaningful part of the conversation. So that was my day job.

But I was the only software developer in a company of architects. And so, I didn't see much of a progression for myself there after a couple of years. But I was very fond of the company and they were very generous to me so what I ended up doing was finding a Master's and potentially a Doctorate program at UCL that I could use to get better at software and to get better at software for architecture. And that was where I met you, at the UCL Architecture School.

We met at the Bartlett School of Graduate Studies, where there was this collection of people who were interested in AI, in simulation, in virtual reality, in mapping, and urban planning—people writing their own custom geographic software and people writing lots of interesting mathematical analysis tools on top of whatever mapping data that they could get their hands on. I think I was interesting to them because I had a working piece of software that could map an airport and show it populated with people. And they were super interesting to me because they were doing that with cities and doing it in 3D and all that good stuff.

So those were the two things I was doing officially before I found OpenStreetMap. I was working for an architecture studio and I was in a Doctorate Program where I met you. And then the third thing I was doing, and I think probably the thing that occupied equal amounts of my head space at the time, was figuring out this whole web-based visualization thing. I was looking at tools to make graphics work in web browsers and playing around with software like Processing from MIT, playing around with the early canvas graphics stuff in the browser, playing around with VRML for whatever payoff I got from that, right? But playing around with how to represent the stuff online and really throwing myself deep into the culture of the internet and how you wrote software for hundreds and thousands and millions of people and got them to communicate with that. I think, I look back now, some of that was incredibly naïve and if I'd known what I know now—how much more effective at that I would have been. I think the overlap of those three things kind of put me in the perfect sweet spot to be captivated by OpenStreetMap and the potential of it.

What was your very first interaction with OSM?

You know, I don't really remember the very, very first stuff. I remember the first couple of times I met you at the Bartlett and I remember ranting about OpenStreetMap on the mailing lists very soon after finding out about it. I remember ranting about OpenStreetMap in bars. I remember sketching what I thought the OpenStreetMap homepage should be with you and Mikel in a pub. But I don't remember my first interactions with the software too much. I remember you explaining to me, it was a Java app that was a web server. I remember one of the early hackathons we had at the Bartlett where Ben Gimpert was showing us how to optimize MySQL queries. And I was already out of my depth with web apps at that point. I'd written some janky ASP stuff at work.

Tom Carden

What happened at first hacking party?

I remember, it was either post Google Maps or Google Maps was very much in the air because one of the things we thought we could do was write a tile mapping viewer in an afternoon. So we were like, "Well, we have this thing. It's rendering images on the server. It would be really nice to use this AJAX stuff that seems to be getting popular. Let's do a Web 2.0 version of OpenStreetMap. Let's have it all run in the browser." And we got some way there, hacking together bits of Anselm Hook's CivicMaps, JavaScript and various things like that. And this is where it's a blur. I'm not sure that was all at the first one but I do remember sitting down at the Bartlett and trying to understand how bits of maps could be stitched together in the browser, both geometry and early map tile ideas, and failing miserably. I think I still have those demos somewhere.

And what about the sketching out how the design should look?

Yeah, I think you and I had met up a few times to talk about various different ideas we had and things that would end up looking like Upcoming.org or today's Eventbrite, right? Or whatever we were interested in, event management stuff and things like that. And there was a general template in my head for websites that were set up for participation, from things like Flickr or del.icio.us and those sites that were just getting popular. So there was a sense of there's some hybrid of those that would work for OpenStreetMap – Flickr was very consumer-ey, very friendly. It had a very clear voice and it kind of talked you through what it meant to be on Flickr and how to be a Flickr person.

So that's something I remember going forward with OpenStreetMap where it should be like that. And then, obviously it should be like Wikipedia because that was the early success model for the kind of collaborative editing we were interested in. And I'd spend a lot time in the years up to OpenStreetMap on a site called Everything2 which was a kind of Wikipedia in the sense that it was also a collaborative encyclopedia. But the model, the community, and the culture of the site were all different. And so, the nuances there around like "Should it be more like a Wiki? Or more like a blog? Or even: should you need to log in before you can edit?" All these kinds of things were the kinds of things we chewed on. But yeah, I should've dug out the sketch before this call so it could jog my memory. I think I still have it.

Everything2, so was that a Douglas Adams thing?

Everything2 was not. No, that was H2G2 – the BBC sponsored or adopted it after Douglas Adams started it. It was another similar thing where it's like "Let's have a page about everything ever," right? But with a very different editorial model to Wikipedia. No, Everything2 was a spinoff of Slashdot. The same people, similar kind of software and the difference there was that once you wrote a topic, you'd kind of claimed that post and the community helped you edit that post and made suggestions and stuff but it was your post and it was your writing. And then, depending on the version of Everything2 you were working with, there was one with a limit of two posts about any topic, and another where there wasn't. I can't remember which one was which. But anyway, that was why it was somewhat flawed. There wasn't one version of the world that was being steadily improved and tweaked by everybody à la OpenStreetMap (literally a version of the world) or Wikipedia (one version of knowledge) being tweaked by multiple people. And the backchannel was hidden as well. So if you think about the backchannel on Wikipedia being very public, right? Like the document itself might be the only thing that most people read but most contributors on Wikipedia are discussing and arguing behind the scenes but publicly.

And then I think all those kinds of threads were the threads we were pulling on with OSM. What does discourse look like when it's about space and how do you find the people who are like you, that are near you? How do you have a local community of mappers when it's a continuous space? But at the same time, how do you have a low barrier to entry so you can kind of get in and contribute, right? Like, those are the early conversations I remember. But also, just a lot of convincing people that it was a good idea.

I think like ten years on, when we see what people are doing with OSM it's like "Oh yeah, obviously, you would have a collaborative map of the world. And obviously, it would be relatively complete in all the interesting places that you want to go to. And obviously, there would be extremely high-quality cartographic renderings of it from multiple different people"—no, not obviously, right? Those initial conditions that you set and the others set with your help—and increasingly with your hands-off "please do whatever you think is right" approach to OpenStreetMap—those things paid off in some ways because it was the system that resisted control the most. Because if it was set up by a corporation, it would have never been open enough, right?

Right, that makes a lot of sense. You know, it's funny. I was talking to Mikel. He's like, "Why did it take off?" Nobody knows.

I'm not the person even to claim 20/20 hindsight but I remember making lots of deliberate design decisions and watching you make lots of deliberate design decisions. I don't just mean design of software. I mean, design of who's in charge? Design of how do you socialize this idea, right?

OpenStreetMap is the meme and I think, for a long time, that was more successful for the dataset or the software. But the dataset and the software filled in, in the wake of the meme. And I think that was somewhat deliberate even if it may not have been optimal. We might be able to go back in designing a better initial set of conditions now, if you had a time machine.

I'm not claiming it was perfect. I'm not claiming it wasn't very lucky but there were deliberate design decisions that went into that, modeled on what you knew about open source software, what we all knew about the web, what Mikel knew about geographic software, you know, like those things. And in some ways as well, rejections of what other people were doing because we saw other models that were trying to do the same thing and failing for various reasons, right?

So back then you said, "let's not get hung up with the best of today," neither the open Geodata of the time nor the open source software, because none of it's very good yet. If you were to build OpenStreetMap today, you would certainly start with some of the better open geodata/open source software. But back then it was all very specialized and it was all quite esoteric. And it wasn't solving the problem that OpenStreetMap was trying to solve which was getting people to start drawing in a web browser. So, let's just make the easiest possible tool to get people start drawing in a web browser. My hacky Processing attempts and Richard Fairhurst's Flash attempts were to say, "Okay what's the best graphic software in the browser today? Let's use that. If it's Flash, if it's Processing in JavaScript and Java or whatever let's just use it to get people drawing."

Let's talk about what Processing was and how you used it and how the editor worked, that you built?

Processing came out of MIT and it was a programming language and a programming environment combined. So, a text editor and a compiler all bundled together that kind of made Java a bit friendlier for designers and beginner programmers. And it taught programming in a graphics-first way. And so Processing—it's kind of a mixed up thing because it's both a programming language variation on Java. It's this programming environment. And then it's also a movement, I guess. Like, it was a set of people who believed in a way of teaching programming and wanted to make programming accessible to people who didn't usually program. And so, it had a kind of culture of workshopping and sharing and teaching stuff. But fundamentally with graphics first and interaction first.

I picked that up and started doing all kinds of things in it. And it became the kind of the language I thought in fastest and the language I could create things in fastest. So when I saw your early OpenStreetMap editor demos and the very, very first editor and I said, "I think you could render this real time. And I think you could move around it. And I think you could click on a street and type the name in and things like that." That was really all I was going for. I thought, "Okay, if I can get that much going. If I can get an editor where you can draw a line and then write on it, that will be great."

We made a ton of wrong turns there in the way that we modeled the data. For example we said segments were sufficient and not ways (collections of segments) as OpenStreetMap has it now, right? So we made some pretty naïve decisions but with the principle of doing the simplest, dumbest thing that could possibly work at heart. I think they were steps in the right direction and it showed that the bar was pretty low but what you needed to do was show a map and allow people to make that map different if it was wrong. Richard did a much better job with Flash later but it was good fun, proving that it could be done and then actually getting it on the site for a while. In parallel, people were doing JOSM, which obviously was the frontrunner for a desktop-based editor and was much more sophisticated. But its barrier to entry was a lot higher too.

Around this same sort of time, do you remember Dorkbot?

Getting folks at Dorkbot excited about OpenStreetMap, I think, was key. Dorkbot being a kind of global movement of "people doing strange things with electricity." And some of them were doing extremely dangerous, violent things with electricity and trying to make musical instruments out of lightning and all kinds of stuff.

After attending for a while we showed up and we're like "We have these maps" and we somehow got an early crowd of excited people. And we weren't the only mappers there. People like Schuyler Erle and Jo Walsh were there doing their things as well and getting a movement together of Londoners who were interested in open data, open WiFi, open maps and stuff like that. So, when we started going regularly, we fit right in. I would show early Processing apps and demos I was doing like the travel time map of London I put together in late 2005.

You and I did a print based on some early data we got as part of the OpenStreetMap project and demoed that, to much acclaim. I think most of the ones that were sold probably went to people that were at Dorkbot London or friends of people who were at Dorkbot London. That was a good place to start socializing those ideas and things like the early OpenStreetMap mapping parties could be announced at Dorkbot and bootstrapped from that existing community.

Do you think it'd be fair to call it a maker meetup before O'Reilly decided that we were all makers?

Yeah, they weren't branded at that time. Or it's like if there was a Burning Man in London, that would've been the epicenter of the people who were doing that kind of stuff. I think every city's Dorkbot had a different flavor. I remember going to one in Belgium and one in San Francisco and they were similar but different. But the London one was kind of hackers and makers as we would call them now.

Do you want to talk about the poster and what we did?

I managed to find it in all my stuff in England. So, I guess, early days in OpenStreetMap, we really believed that GPS traces were going to be a key piece of raw material. And finding industrial sources of GPS traces was a kind of puzzle that we went through for a while. We're thought, "I wonder if anybody already has GPS collection at scale." And saying, "Maybe there are courier companies or taxi companies or maybe the rubbish collection people have GPS on things? Or maybe the buses have GPS on them?" Maybe we could tap into something that's passively mapping the streets all the time as well as each individual OpenStreetMap contributor uploading their stuff from time to time.

There were a couple of things about the print. Again, it was like OpenStreetMap, the meme, was still the thing that we were promoting most effectively rather than OpenStreetMap the dataset or OpenStreetMap the software. Like this compelling idea that you could get a map out of just GPS. You could say, "Here's a bunch of GPS from people in London and it clearly marks the major streets with more emphasis than the minor streets. It clearly covers the vast majority of streets. You can see the parks. You can see pathways through the parks." So that was really compelling.

I think the interesting thing for me is we kind of mumbled through where the data came from at the time. So we said, this is OpenStreetMap GPS data – which it was. It was data that the OpenStreetMap project had but it was very specifically data from a courier company that we had made friends with and not necessarily from individual contributors of OpenStreetMap. I'm not sure that we made that entirely clear at the time. I don't think it really matters but it's an interesting thing to look back at because the poster is really two weeks of courier data rather than OpenStreetMap contributors at the time - except that the courier company was an OpenStreetMap contributor, so. Anyway, you know, fine print and all that.

But yeah, I think at the time, saying "Let's throw all of this data and only this data onto the screen first and then onto a big print and then let's sell that." Nowadays, we do it via Kickstarter as you've done, and try and raise the money ahead of time and see what the interest is. But, we went ahead and kind of invested a little bit of our own money, a couple of hundred quid, got them printed and then asked who was interested. And, of course, massively underpriced it, not knowing at the time that £10 for a thing that lots of people find objectively beautiful was probably a bit cheap, especially by the time you'd mailed them all out—one one by one, in smelly poster tubes and lugged them all to the post office and everything. A bit naïve but I think we did it right. I thought the model was correct.

I remember being halfway through signing them and you saying, "Of course, you're not using your check signature, are you Steve?" Or something like that.

Yeah, you used your art signature which, of course, we all have our art signature down before we start signing those posters. I think my signature now is probably significantly different. So that was the early days of how do you get attention for this stuff. We're like "Maybe boingboing will link to this and maybe a few thousand people will look at OpenStreetMap next week. Or maybe Slashdot will link to it or something." But even just like marking a moment in time where the OpenStreetMap GPS could be enough to trace the map. I think that was huge. This is before OpenStreetMap had any aerial imagery at its fingertips other than odd bits and pieces or the 15-meter Landsat stuff that we were just starting to hook up in there at the same time, right?

Exactly.

It's like "tracing what?," right? In 2005, OpenStreetMap is people tracing maps in thin air basically or on top of GPS data if they had it. So GPS traces are a pretty vital part of the story but easily overlooked now because it's *so obvious* that you would just trace free abundant 1-meter aerial imagery from the lowest bidder (which is amazing).

Do you remember the first anniversary?

I think I do. I certainly remember OpenStreetMap being a very small community for the first year or two and realizing that if you were a loud voice on the mailing list, as I was, that most of the people were just listening.

So I think the first anniversary party was essentially you, me and Ben Gimpert and Alex.

There was the general idea that every revolution needed three characters and it needed a bonafide genius which I'm going to flatter you and say it was you at the time. In the OpenStreetMap story, you were the creative force behind OpenStreetMap. But it needed someone in fine standing in the community to point to you and say, "This guy is not crazy. This is something worth doing." And then I can't for the life of me remember what the third one was but anyway, I thought of myself as the person in good standing who just had to follow you around and say "He knows what he's talking about" and OpenStreetMap would happen.

I remember doing a lot of just emphasizing what was deliberate which I think is maybe where Mikel and I differ in our rosy historical views of OpenStreetMap. I remember a lot of deliberate decisions and I remember defending those decisions fairly emphatically on mailing lists, in pubs, at mapping parties, at hackathons. And obviously, some of those were my own beliefs as well and beliefs in certain ways of organizing collaborative systems and stuff.

Do you want to give some examples?

I think, it gets down to the kind of the thing which is very much in line with the World Wide Web, which is like "good enough is perfect." We would rather have a messy system with a low barrier to entry and worry about cleaning it up later than have a perfect system with no data because it's too difficult to contribute. And so, when it comes to—you know, we may have drawn the line a couple of times in the wrong place. Saying things like "Segments were sufficient and ways could come later," that was probably a mistake.

But saying "Tagging would be sufficient," probably really irritated a lot of people because tags are noisy, disorderly, chaotic and very hard to normalize to 100% clean. But the great thing is that it meant that anybody could start adding anything and that it became a conversational medium in some ways. Like, "Oh somebody's added that. I'll add that as well. I'll add that tag next time I see one of those." That was very important, I think, and easily rejected by people who weren't thinking about a long-term system that would evolve to support many, many different use

cases. That kind of decision was staunchly defended by you and m
other early contributors.

So I think that's one case where there was a deliberate design dec
aligned with maximum potential contributions and it rejected perfection. I think we did that in a number of other places including the software that was being used. Including rejecting geographic databases and rejecting geographic projections in the early days and just saying "we'll get to that." That's not the essence of this. The essence of this is "anybody can get in and collaborate." So, I do have rose-tinted history glasses on but that I remember very clearly.

And it created a forum for community to flower because "hey, all you guys who want to debate stuff, here's the mailing list and here's something great you can debate for the next ten years."

Absolutely, yeah. And instead of saying, "Okay, I have a new way of modeling phone booths in airports, how do I modify the database to support that? How do I get someone to code my change for me," right? It was like you don't even have to ask permission and just start adding phonebooths in airports. And once you've added a few dozen of them, tell people the scheme you've used to mark those up and add it to the Wiki and maybe other people who think like you will start adding those. And maybe, before you know it, OpenStreetMap has the best collection of phonebooths-in-airports data in the world.

And that's a toy example but I think cycle mapping was probably one of the very, very early wins for OpenStreetMap where it was Andy and friends doing that stuff, almost spinning off a fork of the project because they were so prolific with the cycle data, doing one of the first proofs-of-concept of a specific purpose-built front-end to OpenStreetMap that emphasized different data. And they had data that nobody else had. And they didn't need to make patches and change database schemas and make new API methods and all that stuff to support that data. I think that was really key.

I'm not sure that people understood that but the more traditional kind of information systems modelers may have rejected that because it was a dirty data set. And the traditional engineers may have rejected that because it was expensive to compute those queries and to filter by that data. And to render that was difficult as well. But those were all problems we could solve later. We can't solve the problem of getting people to contribute data any other way.

Tom Carden

we touched on it a little bit but is there anything you want to talk bout with Schulze & Webb?

They were among our biggest cheerleaders in the London web and blogging community. People we didn't know especially well at that time. We knew of them and we were getting to know them and they were getting to know us. We saw eye-to-eye on a lot of things, especially after a few beers. The general enthusiasm for the project from folks like Schulze & Webb who, at that time, were fairly influential and had a growing network of collaborators in London—both collaborators within their own company, and then the folks that they were working for at Nokia, the BBC and things like that—who were listening to them very intently. And if they happened to mention that they were interested in the OpenStreetMap project, that certainly couldn't have hurt the OpenStreetMap project at that time.

I remember we leaned on Jack in the early days for – just as an example of probably how naïve the two of us were. We went and asked Jack for help with business card design. He was the only graphic designer we knew socially. So like "Okay, we'll go ask Jack what a good business card looks like." But to see his approach to that, I think, was an eye opening thing for me at that time where he just happened to have a collection of a few hundred business cards, some of which he liked and some of which he hated. And he was humble and straightforward enough to take us through those and kind of deconstruct the business card for us. I think that approach to design has served him well over the last ten years for sure. But I think being that accessible has served him equally well.

Do you think that it was not just the naïveté, it's also the willingness to go ask people stupid questions as part of the beautiful thing here?

Yeah, I think so. I think that's right. We were saying like, "Okay, we know there are people out there who know a lot more than we do about certain things. Let's find them. Let's see if they're interested in this idea. Let's see if they have an afternoon, or an hour to help us, or can point us in the right direction." I think we were lucky enough to have several dozen people like that and two, three or four of them who stuck with it forever like Mikel. And some who were kind of passing characters who nevertheless made a long-term impact on the project like Ben Russell.

Right, so maybe I'm wrong in this but I remember being in—I think it was like Schulze & Webb's office or something like that and there's a map of the U.K. and that's how the mapping party came about. Because there's a map, I said, we should all get together. I think it was

you and I looking at a map and the obvious thing was the Isle of Wight. For some reason, it just seems to jump out from looking at the map of the British Isles. I'm pretty sure that you named mapping parties as well. Do you remember any of that?

I don't. I remember coming up with the ideas but I don't remember if it was me or if it was you or if it was somebody else. But again, I remember very quickly that we came up with a kind of way to draw a ring around what a mapping party would be and what it was for and, even then, understanding that it was essentially a mini-conference. That yes, there would likely be a lot of mapping data generated and that we would likely have time to come up with better ideas for how to model things, and we might have time to patch some software.

But really, we would get to know each other. By going to one specific place and bringing as many people as possible who knew about OpenStreetMap to that place, we would socialize the idea of OpenStreetMap in that place. So if you picked somewhere like the Isle of Wight, that happens to be bounded geographically, you have the potential to map the whole area and say, "We have mapped the whole island." I think that was the key idea at the time. It was to say like, "Let's concentrate all of our efforts in one place and see what that looks like" and how quickly we can advance the State of the Map—as the conference has become known - how quickly we can advance the state of the map in one day or one weekend, but also, as an excuse to actually get face-to-face when the community is distributed online. And, at that point, which is still really early in the community, we captured the imagination of folks who bundled into a van from Norway or flew in from Germany or whatever. That was pretty incredible to see.

What do you remember of the first mapping party?

I don't remember too much about the lead up to it but I do remember bits and pieces from the time on the island and getting there and just the excitement of meeting face to face with people that we hadn't necessarily ever put names to faces for before. And still, it being pre-smartphone, essentially. Well, some people had some fancy phones but rarely did people have a phone and a GPS device and a sufficiently good data connection that they could do any mapping on the fly. So all of the effort was around getting people familiar with, and up and running with dedicated GPS trackers that had limited storage on them but that could record the breadcrumbs as you called them—the Hansel and Gretel trails of wherever you went.

And so people were coming up with various ad hoc systems to record trails and make notes of times and time stamps and way points and things, often on paper, and kind of having these parallel data sets of like "Okay, my Garmin spat out these waypoints that are somewhat accurate because I taped it onto the top of the car. Or I wore it on top of my bike helmet or something like that." Still not brilliant devices and terribly nuanced things to get up and running reliably and predictably. But yeah, having these trails from the Garmin devices or whatever and then having these sheets of handwritten notes at the same time and then going back and sitting at the laptop and trying to figure out when you upload the data whether it was rendered in the right place and whether it had recorded where you think you went. And then trying our best to communicate to people that it really wasn't part of the project to compare OpenStreetMap to another map. And so if you needed to refer to another existing map to complete your mapping that you hadn't gathered enough information yourself.

Having this "clean room" approach to the map, I think that's another bit of the culture of OpenStreetMap that we kind of staunchly defended and borderline ranted about in the early days, knowing full well that the commercial mapping community in the U.K.—the Ordnance Survey and other bigger publishers and owners of mapping database—really wouldn't tolerate any copying of their data. So, chatting with people at the mapping parties about how important it was that we had a stand-alone system.

And a defensible stand-alone system. So we kept the raw data that we collected. We scanned our notes. If we used audio, then we kept a copy of that, and we thought about ways to get that into the OpenStreetMap database in the future or get pictures of house numbers or street signs or whatever and get those geo-coded and uploaded to openstreetmap.org even though we didn't have that capability at the time—talking about collecting that level of data so that there was never a need to refer to an existing printed map.

And I think that's the stuff that got people thinking both "Oh, this is really interesting. These guys are really serious." But also, "These guys are crazy." Like that can't be right. I know what that street is called, can't I just make sure I've got the right spelling by looking at my A to Z or whatever." It's like "Well no, you can't." And so, that side, I remember a lot of those conversations on the Isle of Wight. People knew that but they didn't know how important it was to you and to me and to the other folks who were really anchoring the discussions at that time that people got it through. It's almost "clean room" to the point of absurdity.

I mean, aside from the fact that the OpenStreetMap project has been a successful way to gather data. I think it was also a successful rhetorical device to show people how absurdly locked up mapping data was in the U.K. and Europe. And in some ways, that still might be its biggest success. And that's part of the bigger meme of OpenStreetMap, I think, that still lives on. People like Mike Migurski, in the U.S., are still fascinated by those early aspects of the community. And coming out of a country where there were amazing maps available but the data couldn't be used in creative ways because it was so expensive.

We've talked a lot about how you evangelized OSM and you influenced it. How did OSM influence you?

Yeah. That's an interesting one. So, if I think about the first few weeks I spent at Stamen Design—so towards the end of 2006, I kind of dropped my doctorate aspirations at UCL and parted ways with YRM—the architecture studio I was working with, and decided to throw my lot in with Stamen Design in San Francisco. They were very generous and helped me get my visa and moved me out even though the visualization consulting business that they had was extremely modest compared to where they are today and compared to the number of people who have a similar-looking business today. So, kind of, just to start to tell a little bit of the Stamen Design story.

This is November 2006. I flew out, started working with Stamen. They had shown that there was an interest in visualization for the web and that there were companies willing to pay for it and that they could build up a company around it. I think that was what was interesting to me was that I knew of probably a few dozen individuals and artists and things like that who were doing this kind of work but Stamen was the first company that I saw where you could call them and hire them.

I got there and they were working with an organization coming out of the State of Indiana that was mapping 911 calls. And I remember suggesting that we would use Mapnik to make a base map. Mapnik was Artem Pavlenko's mapping software that the OpenStreetMap project had just started using. It was still a very difficult piece of software to compile and run and deploy at scale. It required a lot of baby sitting. I think, in hindsight, it was outside of my expertise.

It was exactly the right kind of challenge for Stamen Design to say "Okay, we have a little budget." The State of Indiana has lots of open data luckily because it was in the U.S. We could get state lines and county lines and highways and things like that from the Census. And so Tom shows up in San Francisco and manages to squeeze into this project the idea that we would have a custom base map. So that was kind of my very first contribution to mapping at Stamen and to getting Stamen familiar with the open mapping data that was out there and the open mapping software that the OpenStreetMap project had kind of validated.

I think I mentioned before, there was lot of open mapping software out there. Mapnik, I think, was a new generation of stuff that was as good or better than the commercial offerings and written in modern languages, for better or worse. It was written in fairly modern C++ and nicely abstracted so it could talk to all the kinds of different datasets that we were interested in working with. I forgot what our initial theme was for this ramble but—

How did OSM influence you?

I think OSM, in some ways, influenced me by being that like "What's the simplest thing that could possibly work?" In the case of that first project with Stamen, wanting a base map, wanting a custom base map and Mike Migurski had written open source tiling software in ActionScript2 at the time but we didn't have any tiles that we could use for a commercial project. I think that was the interesting thing. So he'd written this thing that could show tiles from Google and tiles from Yahoo and so on.

I think the big influence from OpenStreetMap on me was the realization that it might be *possible* to use those tiles and that they were *just* images and it was *just* software but I was the one saying, "Hey, you know, that data is copyrighted. We don't have an agreement with them. They may not notice us for a while but we should have a 'clean room' approach to mapping." So right from the beginning of my almost four years at Stamen, I was one of the people saying, "We need to be able to do this from scratch."

And shortly after that, fast forward maybe six months or a year, Shawn and Mike from Stamen were presenting at O'Reilly Conferences, a workshop called Maps from Scratch which was showing that both from OpenStreetMap and also from all of the open data that was out there, especially with the U.S.-centric perspective, how you could get maps that were comparable to Google Maps. But also, how you could get maps that were radically different, either artistically, geographically or just from a dataset perspective kind of different looking and doing that kind of visualization-provocation style thing that Stamen does so well. But yeah, my influence from OpenStreetMap was that right at the beginning saying like, "Yeah, we have to be able to bootstrap these projects. We need open software and open data to do this properly. Otherwise, we'll just be dependent on whatever company is in favor at the time."

You transitioned fairly quickly from being deeply involved in the core of OSM to going to San Francisco and then being much more an end-user?

That's right. I describe myself charitably as an "early participant" in the OpenStreetMap project. I was never much of a contributor of data. Like I said, I was a contributor of furthering the meme. I was a contributor of early proofs of concept for various bits of software that were needed. I was a contributor to some of the early kind of social glue and referrals and things like that but I could never claim credit for being an active contributor to the map itself which always stings a bit if I own up to that.

I was always more interested in what you could do with the data. I know that some people would see that's the kind of an uneven thing but I think there were people that wanted to build on top of the map. There were people that wanted to build the map itself. I think that's fine, personally. I think, I found other people at Stamen that were more interested in showing what could be done with the data when the data was free and open. Then we did, necessarily, need people who wanted to contribute to the data.

Although I think Mike kind of tipped the scales the other way with projects like Walking Papers that really allowed people to build up much more sophisticated work flows around capturing original data and feeding it into OpenStreetMap. So I think we made it right in the end. But for a while, we were definitely just consumers of the data and really interested in showing off when there were no restrictions on what could be done with it—apart from credit, of course.

Do you remember O'Reilly Foo Camp, Where and WhereCamp?

There was Foo Camp EU and there was Reboot in Copenhagen. There were a few conferences that year, before I headed out to the U.S. And then I think it was only a year later that you followed and came out as well.

I think those conferences were the start of kind two things. One was again the opportunity to spread the meme—to get to know more people, to meet people face to face who were starting local mapping communities. All of that stuff. For you to get good at speaking about OpenStreetMap as you did hundreds of times in the end and still do. And then also, the start of socializing the idea of OpenStreetMap with folks in the U.S. who were visiting for those conferences. And I think both of those things were key.

I don't remember too much else specific about those events though. There were kind of further—there were other venues for the kinds of conversations I was talking about that happened a lot on the Isle of Wight. They're like "No, really, we're going to do this from a "clean room" approach. Yup, we're getting this data from scratch. Yup, we've mostly written our own software. Yes, we've heard of that open source geographic database. No, we're not using it." Like all those questions, right? And all those conversations - I remember those happening again and again at venues like Foo Camp EU and stuff.

I'm just trying to remember what else. I think that was where I really had the first kind of conversations with the mySociety folks about their travel time mapping and comparing notes from my small attempt to do that just for the Tube Network and their ocean-boiling attempt to do it for the whole U.K., and the differing approaches there, like how you present the data and how you interact with the data. So that was at Foo Camp EU. And then that would seed some collaborations with Stamen a year or two later.

In some ways, we are still seeing those play out now. I think they just got a new version of Mapumental out towards the end of last year and that came out of those early discussions. So I think watching those things evolve and be knocked back and forth like a design/engineering tennis match between Stamen and MySociety, that was tremendously good fun and wouldn't have happened without events like Foo Camp EU.

Some people were just always going to be irritated by the noisiness of the OpenStreetMap data, right? I imagine they probably lost sleep over it. But yeah, they just looked like zooming out a long, long way—looking at like "If we were building OpenStreetMap in the image of the early world wide web and then a lot of the push back we were getting was around strictness and expressiveness of the data model. And isn't there a cleaner way to express these relationships and all of that? And maybe it should just be RDF?" It was exactly the same line of reasoning that people were using to critique the web and saying, "Maybe, the web should be more structured."

It's kind of ironic that those conversations are still going on now and they're still the criticism level of the web that it's messy and noisy and yet it trucks on. And so, I think, some of what we recognized in those early discussions, we were lucky in that we had followed those discussions about the web or we had followed to pull on another thread that we haven't touched on yet around the licensing when people said, "Oh, you can't have a viral-style license for OpenStreetMap because nobody will ever contribute." You can say, "Well, actually we've seen this debate play out between Linux and BSD, and Linux is doing fine thank you." Actually, it's somewhat strengthened by that buffer that it has where you can't co-mingle it with commercial operations. You have to have a clean line. Yeah, being able to say "We have seen this pattern and we are making a deliberate decision based on what we believe the trade-offs are and what's worth it and what's not worth it", right?

Years later, you were criticized for picking a viral license and you're still criticized to this day for picking a viral license as if you didn't know what you were doing. And again, that's one of the things I would say you may have been wrong or you may have been lucky or you may have been both wrong and lucky but you were very thoughtful and mindful and deliberate about what you did and why you chose those things. And so, I was one of the people that was there to say, "Look, these tradeoffs are understood. This is a deliberated decision and we understand, we sacrifice this kind of collaboration or that kind of output or that kind of input but we gain in this strength. We gain in these clean lines in this way and the project being defensible in this way."

I'm talking in kind of abstract terms, but when it came to things like "my company would add their data but we can't because we'd have to give it away," there are people who thought that was terrible and there are other people who said "that's fine, we'll get that data some other way eventually because that data is a description of facts about the world and there'll be readily apparent to somebody when they need them and they can put them in OpenStreetMap and it will be fine."

There's something you touched on there that I'd like you to expand upon if you could, it's that, during that sort of time period FOAF (Friend Of A Friend) and also XML were sort of brand new shit-hot technologies?

It was obvious that the OpenStreetMap API would speak XML. What other thing would it speak, right? Which is strange in our world of JSON now. But yeah, it was one of those things, XML was the obvious choice. In 2005, there was really no other format that you would use to model an API than XML. And then it was a question of "would you use your own XML?"—which OpenStreetMap did. Or would you use an existing form of XML or would you mingle different schemas and name-spaces from other people's XML if other people had gone to the trouble of documenting the right way to model something. The theory was that you would have parsing software that would speak various different schemas and be able to kind of blend its knowledge of different schemas together in interesting ways, automatically or semi-automatically.

I've mentioned our naiveté at various point and in some ways our naiveté translated itself into luck because we were either able to ignore, or unwilling, or unable to understand all of the inputs and options that we had. But I think, in kind of rejecting a lot of the existing XML schemas that were out there and in saying we wouldn't worry too much about blending existing geographic schemas and existing database schemas that we would kind of go our own way. I think that that was wise, even if it was lucky, in the sense that looking back now, I see things like FOAF, the one that you mentioned, but plenty of others as well, where schemas that were out there that were documented and that were able to model pretty interesting systems. Like model an entire social network using XML and describe it as FOAF and ... tools will appear and things will happen and you'll have all kinds of interesting software that can all work with those same APIs.

But it was a chicken and egg thing, right? There was no large dataset of social network information described as FOAF. I didn't have a good enough address book to translate into FOAF if I was interested in doing that. And so, it was kind of analogous to the way we were describing OpenStreetMap. And since when there was no OpenStreetMap dataset, it would have been a waste of time to describe an elaborate way to describe mapping data. And so I think, sadly, a lot of the effort that went into describing interesting open data schemas in XML was a waste of time because there was no data to describe. And then by the time there were people that might have been interested in describing the data that way, the schemas were elaborate and inscrutable and difficult to apply and didn't quit fit with people's way of thinking or way of modeling the world.

I think it's an interesting time in 2005 because you could have spent a long time coming up with an XML schema to model open mapping data. Or you could come up with a minimal one that might have been too minimal but you could put it out there and people like me could write quick Java apps to render it and people like Richard Fairhurst could write Flash apps to render it and you could port a Java app to a Rails app and have it output the same style of XML and have it not be a big deal because it wasn't trying to bring in lots of conflicts, name space support and third party databases and lots of libraries. And it certainly wasn't doing any of the scary standardized API stuff like SOAP and WSDL files and all that stuff.

So I think like we took XML because you had to, but we took very little of it. It is essentially a tree format. We used that and that looked good. And then, either through kind of pigheadedness, willful ignorance, luck or strong opinions or some combination of all of those, we managed to avoid getting roped into most of the schema discussions through virtue of picking formats like tags influenced by things like Flickr and del.icio.us at the time.

I remember reading the Web Feature Service Specification. It was an 80-page PDF and just thinking , you know, "Why on Earth would we ever use this?" The way I saw it was; are you going to put people first or computers first? Because if you're going to convince volunteers to write software against this and to edit data, it had to be super, super simple and very open whereas WFS seemed to have a bunch of different goals to that.

Right and the interpretation was always like, "Okay, you'll drag and drop this file into your IDE." And you go, "Wait, what IDE?" Okay, I might be a Java developer and I might be using Eclipse. And it may happen to be able to code-gen something from a WSDL file or whatever but half of the people who are interested in working with my system are writing PERL or writing Ruby or writing Python or they're writing C or they're writing C++.

The kind of realization that this interpretation was not for open source communities and it wasn't for amateur developers. In the nicest possible way, by amateur, I mean like me and you essentially, where at the time even though both of us were professional programmers, in some sense our attitudes to software were more amateur, more consumer focused - what can I get done in my evenings and weekends, right? Like can I script it? Will it work for free on a PHP web host that's costing me £5-a-month. What will happen if I drag and drop this file onto my IDE and it code-gens a thousand Java files, C# files or whatever because they're just a weird alien landscape really.

I'm pretty sure it was at Copenhagen airport after Reboot and I remember Tom Armitage being there. And I asked, "What should I improve in my presentation?" And he's like, "Steve, don't use the default font." And now, whenever I do a presentation, I always have like Tom Armitage's face looking at me saying, "Steve, don't use the default font."

Yeah. I mean, when you talk about cultures of open source, OpenStreetMap's an interesting one, right? Because it has cultures of open source. It has cultures of software and mailing lists and Linux roots and all of that stuff. But then very quickly, it had very strong academic influences from various corners and especially with cartography.

And it had its design influences from folks like Schulze & Webb. At least at arm's length through me and you and then from folks like Tom that were able to distill design advice into something very pithy, right? He's not telling you which font to use, just the main thing is not the default one because it shows you've thought about it. And depending on what audience you're trying to reach, it will make a difference to how they perceive what you're talking about which I think was the point that you took from him.

It's not like, "Oh, this will make my presentation better." It's like, "Oh, if that's a distraction to N people in the room, I don't want to distract them with that." Or "I'll change the font so they think I've thought about it and then we'll talk about OpenStreetMap," right? And just watching your message evolve to whomever you were talking to, not at all pandering to an audience and saying "What do they need" or anything like that but just becoming a more sophisticated communicator around "How do I get this idea across? How do I get people invested in this? How do I speak their language?" – essentially.

How was OpenStreetMap used at Stamen?

In terms of OpenStreetMap at Stamen, I think it's something like I was alluding to earlier, that we all understood why an open dataset was important. And we were first in line as designers and as visualization people. As producers of visual artifacts on the web, we were first in line to use that data if we could. And so, it made its way into a bunch of different projects that we did. And even when it wasn't really ready for primetime, we would use it to prove that it could be one day used in a certain way.

I think we just kept stepping that up more and more as we worked. Aaron Straup Cope joined us after he left Flickr and made what was called Pretty Maps. And then shortly after I stepped away, Stamen put out their own cartography around Toner and the Watercolor Maps and things like that. That was a crescendo that was just building the whole time to say like, "Okay, if and when this data is ready, we're going to be poised to be the first people to use it for something important. To use it for something visually provocative or just utilitarian and everything in between."

So we were always looking for ways to work it into projects even if it wasn't ready at the time. So like we did the Oakland Crime Maps that we did I think in 2007-2008. We always had a version of that that worked with OpenStreetMap data behind it. In the end, I think, what we shipped always used Bing maps and—actually, no. For a while then, it used CloudMade maps. So it did use OpenStreetMap via CloudMade. And I think that CloudMade was able to bridge that last mile, have the right resources and the right emphasis to bridge that last mile more effectively than Stamen could alone in the late 2000s. Stamen knew what it would look like to do that but couldn't quite commit to the level of server maintenance and on-going updates from the OpenStreetMap database that a commercial offering could. And so, there were early partnerships with companies like CloudMade while you were there and even after you were gone, to do the base mapping for projects like Crimespotting.

Why I say the constant thing, about Stamen, is that they know exactly what you can't do with someone's map if it's not yours. Which sounds like it's a way to describe a hole, right? Like, you can look at everything that Google Maps does and how fantastic it is and how they've executed to precision over the course of years and built up an intensely complex and comprehensive ecosystem of mapping tools and mapping data. I've nothing but admiration for that and yet, at the same time, from Stamen's perspective and even now from my perspective as a visualization professional inside of a large startup, Google Maps has almost nothing to offer me because it doesn't give me the control that I need to say what I want to say.

And think about my medium as a visualization person. I'm a communications designer essentially and I want to control every single pixel that I put onto the screen. And so, if I use someone else's map, I can't do that. If I use open data, I can control the appearance of it but I don't necessarily—you know, what I gain in control, I sacrifice in accuracy potentially. But in the long-term, those accuracy differences drift away and I build and build and build on that level of control and precision I have in crafting my message.

And so, that's what Stamen has been doing for the almost ten years I've known them, is building up the sophistication in what it is they want to say with the data. And like I said, being first in line with whoever has the best platform for expressing open data. So if CloudMade had the best platform for working with OpenStreetMap data and presenting it visually, Stamen were naturally the best partners for them to get the best out of that data at the time. And even MapBox who arguably are the same as Stamen in the sense that they are first in line to be the ones to capitalize on the open data and to craft the message finally. And they have amazing designers and amazing cartographers. They can still benefit from partnerships with Stamen as can anybody else.

And I think I mentioned this about Stamen before, that their specialism is that provocation with mapping data. So finding a thread and pulling on it, that nobody else is finding. And, you know, that's Eric Rodenbeck's kind of impatient, gleeful approach to mapmaking, no matter who he partners with. And as he continues to attract new thoughtful, talented mapping data experts like Seth Fitzsimmons in the wake of myself and Mike Migurski and Shawn Allen stepping away over the last few years. There's always going to be something Stamen has to offer that nobody else has because they've been doing it faster and louder than anybody else for the longer.

Last question, what should we have talked about that we haven't talked about?

We touched only in passing on CloudMade. I think, if we go back to the very, very early days of CloudMade and CloudMade the company that it's become and CloudMade the company that was after you and Nick secured funding for it, and CloudMade the company that became when you and Nick took it on and built a real business around consulting and all that stuff, that was none of my business. But the CloudMade, before it was called CloudMade—me, you and Ben got together and founded. I think there's some interesting stories there that we could talk about if we have time.

We mentioned me and you going and asking Jack about how to put together an effective business card. In those early days where we're like–I think, me you and Ben all got along tremendously well, all saw eye-to-eye on certain ideals and kind of got to know each other through OpenStreetMap and we were like, "We should have a company." We should do professional stuff.

And then between the three of us, I think we all came at things from completely different angles. I remember a lot of agonizing over how we would arrange things. I remember forming a company, putting the money in, getting the limited company together and that being exciting. And then, I remember from my point of view kind of paralysis or I was like, "Well, I already have this day job and I already have this doctorate that I'm trying to do. Exactly with what time I was expecting to build a company?" I don't know, in hindsight. But at the time, I remember thinking, "Well, Steve will get it off the ground." And remember thinking that that was completely reasonable. And then thinking, "And Ben knows what he's talking about because he works in finance. He has money and he has lots of time and he has lots of experience writing software, so he'll be able to help. And sooner or later, the three of us will kind of bootstrap this company." That's my summary: essentially Steve will take all the risk and do all the work. Me and Ben will tell him everything's great. And we'll make it look like we're a real company until we can be.

And some of that I think is just how companies happen. And there are companies that do things that way but I think it was very idealistic of me especially to think that I would essentially take no risks and end up with a company at the end. I think about that like how that time went where we were pitching for work with various folks, trying to get eCourier interested in working with us and trying to get GetMapping interested in working with us.

In some ways it's interesting, GetMapping being the first in a long line of existing and new mapping companies that saw the OpenStreetMap project and tried to control it and thought that by hiring you, they would be able to control it. It's an interesting pattern. It's amazing that those companies didn't talk to each other and figure out "It wouldn't work." I think maybe they've worked out—that getting you to work on stuff that's interesting to you is going to be more valuable than getting you to work on the OpenStreetMap community to try and control it.

Those early days of the company that came to be CloudMade, I think were pivotal for you because you quit your job and figured out how to be self-sufficient and a missed opportunity for me and Ben because we didn't do the same. And it could've been—there's probably a very different kind of parallel universe out there where the three of us did do something and whether it worked to—you know, ten years on, whether the three of us were more successful than we are now in our independent ways, I don't know. But that would be an interesting one to play out. But yeah, there's a few interesting stories in there. I can remember thinking that we needed to have all our meetings after 6 o'clock so that me and Ben could be there. That was silly. And we had rotten luck with the accountant we picked.

Kate Chapman

Bio

Kate leads the organization's technology team and strategy. She is recognized internationally as a leader in the domains of open-source geospatial technology and community mapping, and an advocate for open imagery as a public good. Over the past 15 years she's worked on geospatial problems of all kinds, including tracking malaria outbreaks, mapping private residences for emergency response, and even analyzing imaginary items used in geospatial games. Kate strongly believes in the mantra "people before data" which is core to her current work at Cadasta and her previous leadership of the Humanitarian OpenStreetMap Team (HOT) as the organization's Executive Director and co-founder. Previous to her work at Cadasta and HOT, Kate worked on the geospatial sharing portals iMapData and GeoCommons. Kate also serves on the board of the Cadasta Foundation.

Favorite Map?

"My favorite map is a world map in the Berghaus Star Projection, I have a tattoo of it on my calf."

http://www.progonos.com/furuti/MapProj/Normal/ProjInt/ProjStar/Img/mp_Berghaus-s75-a-54-z-20.png

Interview

Why don't we start with who Kate was and what you were doing before you found OpenStreetMap?

Sure, so I have a geography degree and have done a lot of primarily web mapping type of things. I've also worked as a cartographer so I come from a more traditional GIS and geography background. I was working for a company called iMapData and we made a proprietary web GIS system. We sold a lot of data. I felt like the model of continuing to buy data and then sell it didn't really make sense anymore or it was eventually going to be outdated. I discovered OpenStreetMap and started playing with it around 2009 or so.

What was your first impression?

Well, I'm not even sure that 2009 was the first time I ever looked at it. That's when my account was born. But I want to say I looked at it before and didn't really get it, sort of. It was like uh, and I came back in 2009 and began editing. My first impression was; I don't know—I just sort of just did some mapping in my neighborhood to see how it worked. My first edit was this random island in the middle of the Potomac River, between Maryland and Virginia. It was fun and I kept going. And it sort of turned later into almost like a political statement—mapping my neighborhood. The reason is I owned a house in Sterling, Virginia. Our neighborhood, at the time, was relatively inexpensive and the walking trails and everything were free to use. The neighborhood next door, you were supposed to be a resident of, to use their walking trails and playgrounds and stuff. So I started actually mapping both of them – sort of guerilla mapping the other neighborhood because they actually had cameras up in some areas like "you're not supposed to walk on the trail within their neighborhood." Yeah, so part of it was I wanted a map showing that both neighborhoods actually had very similar resources but the one was, you know, where you weren't supposed to go walk down their trails. So that was sort of how I got started. I met Andrew Turner at some point and we did some mapping parties in D.C.

Do you want to describe those?

Yeah. Well, part of it was I was trying to get a job so I really wanted to work specifically for what was FortiusOne at the time. I wanted to work with Sean Gorman and Andrew Turner. So I figured, doing OpenStreetMap and just getting to know them would be probably a good path to getting hired.

I don't remember the very first mapping party we did in D.C. but I remember the first big one. We mapped the National Zoo in Washington, D.C. At that point, there were already a lot of zoos that were mapped. It was a well-known community thing that people went and would map zoos, so that was what we did. We had about 15 people, I think. That was probably the first large D.C. mapping party.

Do you want to describe what happened on the day?

Sure, so we divided up into groups and we mapped the various enclosures. I believe someone mapped the orangutan—there are these rope structures, so they can swing between enclosures. So, someone had mapped that. We had all ages. I think the youngest person there was maybe two or three. And so, he was carrying a GPS unit around for his father. But it was mostly people who worked in some sort of mapping in D.C. that had come together. Well, people were already doing mapping which I think is sometimes different than some of the other groups versus coming maybe from a programmer background or open data or Linux or something.

And so, did your data involve OSM at all?

I think, I might have worked for FortiusOne at that point, which did use OpenStreetMap data. My job, prior to that, I think they were trying to figure out how they could use OSM data and maybe sell it or something. I don't know. They never figured it out. I personally have never been like a huge fan of ShareAlike but that was actually the perfect example of where it stopped them from basically taking the data in, cleaning it up a little, and then selling it to the government or something. And to be honest, I explained ShareAlike to them so that they didn't do that because there was a company that would take a lot of public domain data, polish it up a little and then sell it for big bucks that was a bit annoying.

And what about at FortiusOne?

At FortiusOne, we used OpenStreetMap data. I think, just using it first a default base map but then at one point FortiusOne and Stamen Design came out with a rendering called Acetate which was designed to put other data like choropleth or graduated dot maps - very traditional visualizations you might see in a newspaper or something like that and make the colors look good. And then the OpenStreetMap data was more of a grey map that didn't really pop so that you could combine the two together.

Okay. And so, during this time, how was your relationship with OpenStreetMap changing?

Well, when I first started out, I was just sort of mapping by myself. I don't know if I went on the mailing list. I think I actually joined the OpenStreetMap Foundation pretty early on because it seemed like the thing to do. Though, I didn't really understand what was going on. Other than people in Washington D.C., I had never met anyone else involved in OpenStreetMap. But with FortiusOne, they sent me to State of the Map when it was in Spain and that was the first time I really got to meet the global OpenStreetMap Community. So, at that point, it was partially my job because I spoke at State of the Map about the work we were doing at Fortius One and OSM becoming less of a hobby.

Right. So, I'm trying to find the right jumping point to formal involvement, you know?

Yeah. I would say my first maybe formal involvement would be, we decided we wanted to put on our own State of the Map U.S. and then formed OpenStreetMap US.

That's a great thing to dive into.

Yeah, so what happened was we thought, "Well, not everyone's going to be able to go to Europe or somewhere far away every year." That year in Spain was the year FOSS4G was also in Spain so a lot of people had to pick one conference or another because if you're coming from North America almost no employer will send you to Spain twice in a year. It just won't happen. So we thought, "Well, if we do one in the U.S., it's more reasonable for people to pay their own way or to just do work-travel." We created OpenStreetMap U.S., incorporated it as a non-profit, and started getting members and planning a conference.

When you say 'we', who were those people and how did you find each other?

I suspect I will miss some of the people that originally did this. The people that did the majority of the work for State of the Map U.S. were myself and Thea Aldrich. We did the majority of that work, putting that conference on together, but there was an original board of five people. Because the board changed year to year, I'm trying to remember who the original board was. I believe Serge Wroclawski was one of the original board members. I can't remember who the other two were. It might have been Ian Dees or some of the other people that have been board members over time.

What prompted the set up of the U.S. board?

Well, honestly, we wanted a checking account so that we could run a conference and didn't want it to be in an individual's account so by incorporating as a non-profit—there are a bunch of different non-profit models but the model I knew came from–actually the roller derby movement. I had been involved in setting up the D.C. Roller Girls. Originally, it was an LLC and became a non-profit. I was on their board when it became a non-profit. And so, I knew a little bit about structure and the legal rules for setting one up in D.C. so I said, "Hey, I think I can do the paperwork." Lawyers have since looked at it and had been like, "Oh, good God, why did you do the paperwork yourself?" but at the time we didn't have any startup money.

So then what happened with the conference? What was the planning? And then what was it like—the event and so on?

There was a lot of planning. We didn't have a lot of sponsors. I know that there were at least three of us that did the lowest level of sponsorship as individuals because we were a little short. I believe CloudMade probably made up the difference as well. And we had maybe 60 to 70 people – somewhere in there, I think. And it was at the convention center in Atlanta but it was definitely not a shiny professional conference. Thea and I stayed up most of the night, the night before, making gifts for people and things like that.

I thought it was pretty good.

Yeah, I thought it was good. It had a different feel than maybe other conferences since but I thought it was great to bring people together and start out.

And how did the conference go?

I thought it went really well. I thought it was great to get people from the more casual hobbyist mapper and then see that there were companies interested. There's always been a little push and pull between, let's say, those interests. I think getting everyone together in a room is important. And seeing that people like the

USGS were there. That was the first time Randy Hale ever presented about teaching high school students in Red Bank, Tennessee about OpenStreetMap. And I thought, "you know, see what other people were doing. "

So what happened next?

I don't think we did another conference the next year. I think it skipped for a year or two. For me, personally, what happened next was there was the earthquake in January 2010 in Haiti. I was still working at FortiusOne and what I did is I was actually trying to gather data to put in GeoCommons—this, you know, is a web data-sharing site so that we could share the data with first responders and other people that might need it.

And I noticed that, on IRC, people were chatting about mapping for the earthquake within the first day or two. Because I think the first person that mapped after that earthquake, it was maybe five or six hours later, an individual in Germany and an individual in Japan started mapping without talking to each other, a couple of hours apart. And then other people just kept going. And so, I noticed that that was going on and I started helping and doing what I could. I was based in Washington D.C. so I had certain contacts with for example, in the World Bank. I was working with the Pan American Health Organization and other groups that were becoming interested in that data. Fortunately, my employer let me just start doing OpenStreetMap full time for like a month and they didn't make me do any work-related things related to our product delivery or anything I was supposed to be doing.

And so, about three weeks after the earthquake, I got sent to Miami, to the Southern Command—one of the command centers for the U.S. Military. And the reason we were sent there is partially because of OpenStreetMap but it was also partially the spontaneous volunteering of thousands of people around the world, both doing stuff with data but also writing software to try to help with the response. I went there and, you know, the Coast Guard was using OpenStreetMap data. It was the most up-to-date map of Port-au-Prince after the earthquake. Part of the reason was the National Mapping Agency's building collapsed and it took a while for people to find the backups, to access it and those sorts of things. And also, the terrain had changed quite a bit.

What were they using the data for and how were they using it?

It was primarily just used as a base map for having road information because there were other efforts going on through Ushahidi to collect reports and then essentially put push pins on the map to highlight people who had needs but without having place names and the names of roads in place they couldn't do that. We were able to get road names—because prior to that, people were simply using satellite data and tracing over was—the Texas Map Library which had a lot of old public domain maps so people were taking road names and transcribing them over to OpenStreetMap.

And those people were all around the world?

All around world, yeah. At that point, about 600 people helped over the course of a month.

So, day to day, what were you doing in the few weeks before Miami and then what were you doing day to day when you were there?

I did a couple of things before that were simply making videos on how to do mapping that were specific to mapping in Haiti. It was simply looking at satellite information and showing people how to map a road. There were also internally person camps and those sorts of things that you could see from satellite images. There were all these people that wanted to help but didn't know how to do OpenStreetMap at the time. Though, honestly, the majority of the edits were made by 40 people out of the 600, and 39 of those people were already OpenStreetMap contributors. So at that point, it really was the seasoned/experienced willing-to-map-all-day type of people that were doing the majority of that mapping.

Can you talk about what they were doing, like the tools they were using, the data they were using?

Sure. The videos I made were using Potlatch 2 at the time, I think. Would it have been—maybe Potlatch 1? And so, my videos were just showing a web browser and how you could edit from the web browser. There's a tool, Map Warper, that allows you to take a map—so we were taking the maps from the Texas Map Room and lining them up over the existing OpenStreetMap data so that they matched up, so that you could view them and then add road names.

One of the things that I think really was a game changer is that a lot of satellite imagery was released – both satellite images and aerial photography. The World Bank facilitated airplanes to collect aerial images and they released them under a public license. That's the first time that really, that was available for anyone to use. There were people who weren't mapping where it was, just a massive effort of taking terabytes of data and munging it so that people could use it.

And so, was this particularly different to previous disasters?

There had been some work in 2009. There was a typhoon in the Philippines where MapAction used OpenStreetMap data which I believe is the first official use of OpenStreetMap data in a humanitarian response. I don't think there was a mass effort to do mapping at that point or if there was I wasn't involved in it. I think it was just—

What happened to MapAction?

They're still around. It's actually really difficult to be a member of it. You have to spend one weekend a month in training and you have to go to nine out of twelve trainings a year. And they camp out year round during the training.

Are they all around the world?

No. So that's the thing, you basically have to live in the U.K. to be a member, or be able to get there once a month. And they camp out all seasons. I went there after the last State of the Map that was in the U.K. and was that in September, maybe? We were camping and it was quite a bit cold for someone who was living in Jakarta.

So what did you do?

It's a lot about sort of emergency mapping and techniques for data management and that sort of things. That's the type of training they do.

Is it still relevant with OpenStreetMap?

Oh, definitely. We talk to them regularly.

Okay. How so? So, what's the relationship and how do they work?

Well, so if they're going to deploy somewhere to provide maps, they typically deploy with the United Nations. And so, we're one of the first places they check for availability of data.

Yeah because who else are you going to go to, I guess?

Yeah. Well, there's preparedness exercises to gather government data. And depending on the country, sometimes it's good, sometimes it's not there. So, yeah, they do a data scramble and we're near the top of the list of what they check for.

So did you ever go on the ground in Haiti? Did you fly over or anything?

I went to Haiti twice. The first time was in June. A couple of other groups had been prior to that. I think there had been at least three or four trips of people who were going to help with OpenStreetMap prior to me going.

In June, I went for two weeks. We basically taught people how to do mapping. We would have these huge workshops, or they felt huge at the time - we've had bigger ones, of 20 to 30 people who had really varying technical knowledge. We would teach them how to use a GPS and how to do a survey, to write information down on paper and then how to enter it into OpenStreetMap.

What was the result of that? Did they all start mapping?

Well, so there was already mapping going on at the time. The International Organization on Migration (IOM) actually was paying surveyors to go map in OpenStreetMap. And one of the reasons for that was there was a piece of software called the Displaced Persons Tracking Matrix which was essentially a registry for people that were displaced. And when you registered, those records were tied to OpenStreetMap for their location, tied to like a road segment or a neighborhood – depending on how detailed the information was. So, it was very much a part of what IOM was doing in their registration and management of the IDP camps that were in Haiti at the time.

What's IOM and IDP?

Yeah, so International Organization on Migration is IOM. That's a large international non-governmental organization. And then IDP is an internally displaced person. So, to be a refugee, you have to cross country boundaries. If you're displaced in your own country, you're an internally-displaced person.

When was your formal involvement with the Humanitarian OpenStreetMap Team (HOT)? When did it start? Was it before all of this? Was it during or after? And then how did it happen?

HOT was sort of an idea but there weren't a lot of people gathered around it in early 2010. I know that people had met and discussed it before. I believe Mikel had done some presentations at some point. But when people started going to Haiti, they started calling themselves HOT, basically. I believe – and that was also the point at which Harry Wood registered the domain hot.openstreetmap.org and threw up a wordpress blog so that we could write about what we were doing. And so, people just start blogging. It wasn't anything fancy, just "Hey, this is what we did. We went and trained these people" and put up pictures and things like that.

Okay and then how did it go from there?

While I was there, the first time, we met someone who wanted to give us a check. And so, the way people were going to Haiti before was, we were subcontracting and getting money as individuals through OpenGeo because the World Bank was giving them money to do that. And they said, "Well, this isn't really our business. It doesn't really make sense to do this." And the person we had met said, "I want to write a check to an organization. I want someone I can contract with." And I said, "Okay, I know how to make a non-profit. I had done it the year before with OpenStreetMap—U.S. So we went through a process where we created articles of incorporation and we consulted people, there was a HOT mailing list at the time which still is heavily used, asking people what their opinions were. Some people started chapters of other organizations but there weren't a ton of people participating in that process. So then, at some point, in August of 2010, we actually incorporated.

Soon after, we got this check. It was from IOM. And we went back to Haiti, basically. This time, doing all the organization ourselves.

And so, do you want to talk about that?

About going back or..?

Yeah because I think we covered your first trip but not the second, maybe I'm misunderstanding.

Right. No, that's true. So they were creating these bulletin boards that were kind of like—imagine if you went to summer camp and there was a bulletin board that would give the schedule of activities for the day and maybe there's a map. And so, the idea was to have maps be a part of the bulletin boards so people in these displaced person camps could see what the resources were, know more about their space. And also, work with the mappers we were working with, engage with them and map things that were important to them.

And so, so we were doing that and that's also the point where OpenStreetMap Haiti began to form. There were a lot of people who we had been working with since the first trip. The first trip of people who were interested in OpenStreetMap going to Haiti was in—I think it was around early March. It might have been late February.

At that point in the earthquake response, everyone was still sleeping in tents and there wasn't a lot of room. It was difficult. And finding people to do mapping then versus in August/September when the acute emergency was beginning to be over and people were starting to look a little bit at recovery, so there was time to do things like mapping. At that point, it was when there were so many people interested in mapping. We were doing lots of workshops. But then also, we were forming a mapping team to be more reliable to go into specific displaced person camps and map them. And so, people started talking about creating OpenStreetMap Haiti at that point.

Do you know what OSM Haiti looks like today?

So there are actually two associations in Haiti related to OpenStreetMap. There's Communautaire OpenStreetMap Haiti and then there's one that's just Saint-Marc which is a specific commune in Haiti, so just one district basically. OpenStreetMap Haiti and HOT together did some contracts and did some work together but it was a bit difficult. So we haven't worked closely, on a paid project recently. We still support mappers in Haiti but we haven't worked closely in a business sense since 2011.

Do you want to talk about some of those differences of opinion, both internally and with other organizations? I don't want to just talk about how crappy people are and drag it out but at the same time I don't want

to just pretend it's all rainbows and unicorns, right? And so, if you can talk at any length—internally or externally, you know?

Yeah, so there's a couple different things. So COSMA which is what they call the OpenStreetMap Haiti for short. The structure to become a member to vote for the board of directors, I think you somehow had to get invited in. I'm not quite sure of the process. But one of the difficulties is they were making it very difficult for any new people to join the association.

What kind of challenges have you had with people internally in HOT?

There's definitely differences of opinion on what it means to be HOT versus what it means to be the non-profit that calls itself HOT. Some people think that you should be able to just, say I'm doing humanitarian-related mapping in OpenStreetMap, I'm part of HOT. But, for HOT, our revenue last year was $ 1 million. It will be more than $1 million this year. So from a liability and legal perspective, we have contracts, we have grants, we need to keep up with a certain level of appearance and service and that sort of thing as well.

How was the organization structured? Assume people don't know anything.

Yeah, so our legal structure is we're a non-profit, incorporated in Washington, D.C. We have tax-exempt status from the U.S. government. We're considered a public charity and what that means is if you donate money to us and you pay taxes in the U.S. you can write it off on your taxes. So that's the legal structure but it also means, you know, we can't lobby politicians and things like that. There are certain rules as a public charity you need to follow.

As far as the internal structure, we're a membership organization but you become a member by your volunteer work, or your advocacy, or helping the organization. You get nominated by an existing member. And so the way we initially started was five of us formed the organization. I think we were the initial temporary board and then we nominated new members to sort of kickstart the idea of having a pool. We just actually added about 20 new members today. Today, we're about 64 and I think we've added about 20 more. So, we're up to 80 to 90 members now.

I mean, it's comparable to the OpenStreetMap Foundation (OSMF)?

Yeah but you don't pay money. It's actually, roughly once a year, we have a nomination process. And then people vote to confirm you. And you need two-thirds confirmation from the membership to get in. Not everyone always gets in.

Sometimes it's confusing too, like "Well, that person did all this work. Why didn't you guys vote for them?" So I think there's a difference. We've never been clear on why someone should be a member. It's one of those things where I thought it was good to create a membership organization where paying money wasn't why you got into it.

But I didn't spend a ton of time thinking through what would it look like now that's been four and a half years instead of being 10 or 15 people, because the board of directors is elected out of that membership. And so, you have to be a member to run for the board. It's been okay but there's definitely certain expertise where it can be difficult to get someone into the membership who doesn't map a lot in OpenStreetMap. But for us, mapping in OpenStreetMap isn't necessarily what makes a good member.

Right. So it seems a shame that you're missing out. You mentioned that people who've done a bunch of work don't get voted in.

Yeah. Sometimes, I think there are political and cultural situations going on. There was someone in particular that did a huge amount of advocacy within the U.S. government and I think there's probably some people that it concerns them that he used to work for the U.S. government but he doesn't now.

I think, at some point, we'll need to re-define exactly why someone becomes a member. Personally, if I was going to do it again, I would have half an elected board and half an appointed self-perpetuating board. Meaning, there would still be some form of membership or something. The Wikimedia Foundation has sort of a similar form. It is also tempting just to have a board where you form the board and the board recruits new members. A lot of the non-profits I know do that. There's more stability that way.

Do you want to pick some HOT projects since then, as many as you like, and go into as much detail as you like in what HOT was doing?

Sure. The one I'm personally most proud of is our work in Indonesia. The first time I went to Indonesia was in April or March 2011. The reason I went there there is that the Australian government had heard about the work in Haiti. They knew some people at the World Bank who were involved with us and they said, "Wow, imagine if you mapped things before there was a giant earthquake?" And so they invited myself and Jeff Haack out to Jakarta to spend a week there and essentially talk. I think it was a two-day mapping workshop as well where we mapped one of the neighborhoods out behind their office and spent a week sort of educating people about OpenStreetMap.

At the time, I still had a full-time job. I took holiday time from FortiusOne to go to Jakarta. I'm not sure what Jeff was doing. Jeff's done OpenStreetMap projects for a long time. Prior to that, he had worked in Palestine and he had worked in Georgia with JumpStart doing mapping. And he was available at the time.

After the week, they said, "We'd like you to do a proposal for a pilot." And so, Jeff and I wrote a proposal and it was for a couple of months. The idea was to test out the idea of trying to build an OpenStreetMap community in Indonesia. And so, I went again for two weeks. And then it became clear, I needed to go for months – at least a couple of months. And so, I quit my job at FortiusOne with really only a couple of months of guaranteed money and went to Indonesia.

Jeff and I went to Indonesia. We hired two interns. I gave a workshop at the University of Indonesia and asked if anyone wanted to be an intern. And the professor said, "Here, are these two people. They're great." Little did I know they were actually the only two people who volunteered. Fortunately, they were amazing. It was amazing, we got to travel all over Eastern Indonesia. The place is, you know—most people will never see.

Do you have some memories you want to share?

Well, it was just things like—you know, people go to Bali, for example, on holiday. Some people, it's their once in a lifetime vacation. We would go to Bali and teach OpenStreetMap. That's not the only beautiful island in Indonesia. The country really varies. You'd be driving through jungles and these small towns. And just seeing there's both natural beauty but then there were all these sultans had these palaces that you wouldn't probably go to unless you happen to be in the area.

One of the more memorable experiences was actually one of those things where we're like, "Well, maybe this isn't safe what we're always doing" because we used to go to this small island. It's not even a small island but there's not a lot of flights there. We were taking this airplane and we get off the airplane and we go, "Wow, the engine's really, really charred on that airplane." We took a picture of it. It's just black. And the guys who picked us up, we said, "Hey, have you seen this?" and they go, "Oh yeah, that airplane was on fire last week. It had to make an emergency landing." And so I just never thought when I started doing OpenStreetMap that the type of questions I'd have to ask myself was, "Is it better to take the plane that was on fire last week but the pilot's done a successful emergency landing? Or should we take the one other plane that's never been on fire?"

So, I was just sort of amazed that what had started from mapping in my backyard. It started as a hobby for me and then I'm traveling to all these places. And at that point, we were continuing to do pilot projects. Or we kept extending our contract. So I would go back to the U.S. for a month or two and then I would go back to Indonesia for two months. This was the first year but then I moved there for almost three years.

The thing was there wasn't anything in Indonesian about OpenStreetMap at that point. When we first started doing the workshops, basically our interns translated for us. But they were geography students and they were smart, so after about a week they taught the workshops themselves. But we needed to create good training materials because there wasn't really anything that existed that was really good for repeatedly giving the same workshop to people over and over again who weren't super technical.

Most people knew how to use computers and they were okay but we occasionally—like one of the big problems we'd run into is we'd ask people if they had an e-mail address. So we'd say, "Raise your hand if you have an e-mail address" because they need an e-mail address to sign up for OpenStreetMap and half the room would raise their hand. And then we would say, "Who has Facebook" and almost everyone raises their hand. I'm like, "well, you probably have an e-mail address so you signed up for Facebook" but we were facing challenges that you didn't have for the typical mapper that had mostly used OpenStreetMap before. We had trained people who were uploading their data by putting it on a USB stick and going by motorcycle for four hours to the internet - a very different environment than we were used to.

The majority of people reading this who wind up in Indonesia, what do they need to know about it? What's it like? What happens when you land? What are the differences? I mean, are there Taco Bells everywhere and McDonad's? I mean, you're going to get people who just don't understand?

There are McDonald's everywhere. The statistic I most like to say about Indonesia is it's the fourth most populous country in the world but a lot of people forget that, so it's the fourth most populous country in the world. It's 15,000-odd islands of all varying sizes. Jakarta, the capital, is a megacity. Over 11 or 12 million people live in Jakarta proper but there are 20 million in the metropolitan area. But then, there are very rural places where people are living on less than $1 a day as well. So there's very wealthy and very poor. But then Indonesia is also prone to a lot of disasters. The majority of deaths from the Boxing Day Indian Ocean Tsunami were in Indonesia, on Sumatra. There are volcanoes that go off all the time, severe flooding, and landslides. This country that's developing and its' really become a middle-income country but they still have a lot of natural disasters and problems that I would regularly come into contact with.

Can you talk about how the map changed, sort of, before you were there and then after your three years or so?

Before I was there, there were people who primarily had moved to Singapore to work. And they would map when they came back to visit their family, it seemed like. There were also some entrepreneurs that were interested in not paying Google a bunch of money for Google Maps, for example. But no one was doing a lot of heavy mapping.

One of our goals was to map a million buildings which we accomplished in 2013. There's a lot more details. In Jakarta, the Jakarta government uses

OpenStreetMap as their official base map for reporting flooding. When you look at an official government map saying where it's flooded, that's OpenStreetMap.

One of the projects we had done is there were no neighborhood boundaries in Jakarta. So remember back to we're talking about 12 million people. It's a large province and there's no neighborhood boundaries but they report flooding based on those neighborhood boundaries. So what we did is, we asked people where the boundaries were and we put them into OpenStreetMap.

That was a huge effort. We trained 100 college students from the University of Indonesia to help. And then we actually went and asked the residents. We had five events and we asked the village leaders of these urban villages. We asked them to come and to bring one other person and help us fill out the map basically.

And so, in Jakarta, it's really the most detailed map you can use at this point. It's the only source of those neighborhood boundaries. And now, we're mapping the sub-neighborhood boundaries. And so, if you imagine, there's 267 villages and then there's five to fifteen neighborhoods. And then within each neighborhood, there's five to fifteen sub-neighborhoods—we're talking about a huge amount of boundary information that just doesn't exist at an aggregate level. It's drawn on maps in people's offices.

That's really quite impressive, isn't it?

Yeah. I mean, it's the source of that data. I think one of the other things is—part of the reason we were going there is there's an open source impact modeling software called InaSAFE. It's designed to take scientific hazard data such as a tsunami or an earthquake and combine it with exposure data which would be infrastructure or population data. That way, you can overlay "if this earthquake happened here, what would be the effect on this infrastructure or this population?" And a lot of times, OpenStreetMap is the source of the infrastructure data because if you want to do an InaSAFE analysis and your area you're concerned about, your area of interest doesn't have any data, with OpenStreetMap you can go create that data. So there's been quite a few mapping efforts to create that data to do analysis, to prepare for disasters.

I'm curious, how much of the market plays into this because in the United States, flood insurance is provided by the Federal Government as a monopoly. What that means is whenever there's a flood, the government just pays to build your house again and that's starting to change a little bit. But it means the people are protected from making bad choices, right?

Yeah.

Or they're encouraged to make bad choices. I'm curious, if there's any degree to any of that and stuff you've seen elsewhere.

In Indonesia, specifically, most people are not insured. If a really bad disaster happens, potentially, people's houses get rebuilt through the government or aid agencies but it doesn't always happen. And the people who are most vulnerable aren't like they are here in the U.S. where a lot of times you'll see people had built this crazy house in a place that's really prone to flooding from hurricanes or something like that. In Indonesia, it would be the poor have been pushed out to that place that they shouldn't build. In Jakarta, for example, there's no build zones around some of the river banks but people have built on them. But they're people who really can't afford anything else.

Yeah, so that was going to be my next question. When you say "they were pushed" what were the factors that were pushing them into these risky areas?

I think just that there's not space more than anything.

Okay. So they know it's bad. It's not education. It's not anything like that.

Well, at least in some cases. Let me give you an example, we can forgive the people of Pompeii for building beneath Mount Vesuvius, right? Because they probably didn't, you know? I don't think that they probably had any historical knowledge about volcanic eruptions near them, maybe I'm wrong, right?

Yeah, well, people are funny about it. For example, there was this local government official who became quite famous in Yogyakarta in Indonesia. As the local lore goes, he had taken billions of Rupiah in sponsorship money for an energy drink. He was like the spokesperson for this energy drink. And then the volcano started erupting and, I guess, he said, "This is all my fault. I took all this money for this energy drink" so he didn't evacuate and the volcano erupted and he was killed. If you talk to people say, "Okay, this isn't a good place to be. The volcano will erupt again." But they have, say, "Well, that guy who took all the money from the energy drink is gone so it won't erupt again."

And so, you have this dual look at things where it'll be like, "Well, you understand it's dangerous." And they're like, "Yeah, I do know but"—you know, there's the but. So there are some people where maybe they could move but a lot of times there are people who don't have those resources or they've lived there a long time. A lot of people don't move around. They're close to their family. They've been in the same village for a long time so it can be hard to get people to change.

What are the other projects HOT has been doing that you want to mention? And then, has it ever paid off in the sense that when they said, "Hey, why don't we map before a disaster?" has that paid off yet? Have you had the right disaster yet?

Okay. Well, so we've done workshops in a lot of places. We've done mapping in Mongolia which was the first time we ever did workshops where it was so cold that you couldn't go out for more than an hour to go do mapping. We've supported OpenStreetMap Philippines for a long time. The Filipino mappers, every time there's a typhoon or earthquake there, a lot of times they try to map before the typhoon hits or if there's available satellite imagery, if there isn't base map information. I've been to Vietnam. We were mapping public transit routes in Haiphong. And the idea was to use OpenStreetMap to put the bus routes in OpenStreetMap and the bus stops. We've worked a lot in Francophone Africa to support mappers there.

One of the other things I'm proud of is we recently did flood mapping in Malawi and HOT was able to hire and support one of our team leads from our work in Indonesia to go to Malawi and then Maning Sambale, who is one of the lead OpenStreetMap Philippines mappers, to both go to Malawi and teach people how to do mapping there. I'm excited for it to be, you know, sharing internationally OpenStreetMap and rather it being the very privileged bringing something from the U.K. to people, you know? And it gives people a livelihood as well. HOT has 15 staff.

It sounds almost like Haiti was sort of a turning point. And now, it's just normal when just everyone just uses OSM. I mean, it's nothing special.

Well, I mean, there are things that are special. So I've talked a lot about our going somewhere and teaching people OpenStreetMap. But since Haiti, volunteers from all over the world get together and respond to disasters as well, like OpenStreetMap's the main source of base map data for the Ebola Response. Doctors Without Borders was using OpenStreetMap as their base data to plan, basically. If you don't know what village is there you can't go and see if anyone's sick. And I would say, for a lot of organizations, it is becoming the normal source. It's not a given but I will say that when I say talk about OpenStreetMap and then start to explain it, a lot of times people will interrupt me and say, "Oh yeah, I know what it is" versus in 2010 no one would know what it was when I said that. It's expected.

In November we launched the Missing Maps Project and the goal was in the next few years to map, 20 million of the most vulnerable people in the world. And the partners are the Humanitarian OpenStreetMap Team, the American Red Cross, the British Red Cross and Doctors Without Borders U.K. I mean, HOT's an equal partner with these huge organizations that have a very long history. So I mean it is, in some ways, becoming more normal.

Is there a sense that some of these disasters get a lot more attention and therefore resources than others because they're fashionable in

some way, that they involve westerners or they're close to a western place or something like that?

It's definitely hard sometimes to get people together for a small disaster even if it was really tragic for a small group of people or a really prolonged disaster like some of the humanitarian crises like the Central African Republic. We've been doing mapping for that for, I think, now a year and a half. And it can be a lot harder to get a lot of people together to do that sort of mapping. Maybe if you look to OpenStreetMap effort and compared it to individual donor dollars for those, I suspect you would see some sort of correlation as well because media interest certainly helps.

I will say that we've started doing—now that we partner a lot more with humanitarian organizations through Missing Maps, there is a monthly mapathon in London and Washington, D.C. now. It's usually held at either the British Red Cross or Doctors without Borders when it's in London or at the American Red Cross in D.C. So, they're supporting it. So, that continually gets a lot of people and typically sells out as in there's no more room for people to come help in person.

So HOT's budget, how does it roughly break down? You said, circa $1 million in revenue?

So, half of our money comes from our work in Indonesia. We have a team of 13 there and we have an office that we share with WikiMedia Indonesia and the Web Foundation. And so that goes to a lot of different things. We build software with that budget. We employ staff. We actually have a full-time communications specialist who makes infographics about OpenStreetMap and sends out the newsletter and that sort of stuff, all in Indonesian.

And then the rest of our budget comes from a combination of money to go teach people to map somewhere and software development. For example, we're currently working on a pilot where we're re-launching open aerial map, which is actually a project separate from OpenStreetMap, but the thing is we use so much satellite data and UAV data and we have nowhere to put it. There's no catalog and hosting for open imagery. And so, that's something we're working on because it very much helps enable the mapping that we do.

And so, what do you see? Where's HOT going to go in the future?

Well, one of my major goals is to support a microgrant program. What I mean by that is there's a lot of really interesting OpenStreetMap communities all around the world that start for different reasons and there's a lot of places where the barriers are simple things like they don't have a smartphone or GPS unit, or they need bus fare to go somewhere because they want to map or they don't have internet access—very sort of simple things that are barriers. I'd like to be able to give people, through an application process, $5,000 to $10,000 or equipment to be able to facilitate that because one of the things HOT does now with projects is we often bring our equipment. We have a thing called a HOT kit which it's a waterproof case with a laptop, a printer, cameras, mobile phones and GPS and all the bits to put it together. It's a mobile mapping lab, essentially. And so, we take those places. But to take a HOT kit somewhere a contract and a project are needed. I'd like to be able to provide that equipment outside of specifically doing a major project where we send people to go do training and support more local communities that way.

Does Walking Papers fit in there somewhere?

So that's why we have the printer actually in that kit.

Do you want to talk about—I don't know whether it has or, you know, people keep complaining that Walking Papers is dead but has it been very useful in the past?

Yeah. So beyond the print maps and write on has been key. With Walking Papers where you can scan them and upload them, that actually doesn't happen that often. We've switched over to using Field Papers which was meant to replace Walking Papers. And that's used extensively just because if you hand someone a pen and a paper, as long as they know how to write, you can help them get oriented. If they're not good at reading a map, that's pretty easy to get started with. We're putting a bit of money into Field Papers in the next couple of months to add some features to it as well. One of the major things is it's not localized. It's only in English. And the other thing is you can only print on one size of paper – on a letter-size paper so we want to be able to print larger. So, it's important enough to us that we are putting money into it.

What's the best way for someone reading this to get involved in HOT?

Well, I guess, one of the things to specify is though our membership is about 80 people, we have thousands and thousands of volunteers with various levels of involvement similar to the OpenStreetMap community. If they go to our website, Hot.openstreetmap.org. There's a "get involved" button and there's instructions on if you're a software developer, how you can help. If you just want to map. And so that's really a good launching point.

Okay. Alright. Now, I'm not sure if there's anything in between that and the OpenStreetMap Foundation, is there? I'd like to get you to talk about OSMF but I'm just trying to figure out if there's something that should bridge the two.

I think the first time I ran for the OSMF board was in 2009 maybe or 2010.

What made you decide to do that?

Kate Chapman

I don't remember if I decided to run beforehand but there's this State of the Map where everyone sort of yelled at each other, in Girona? There was not a lot of yelling but there were people yelling during a Q&A or something. And I just thought, "I don't even know who these people are that are yelling at each other." I think you were involved. And maybe Andy Allan. I mean, it was clearly people I know now but as a new person, it was sort of - I thought I could help. I don't know, at the time, I actually had not very much management experience but I thought I could help change things. I thought I could make it more diverse, more welcoming to women and minorities, under-represented groups because it's never been a particularly welcoming place. It's a very typical old school open source attitude. And so, I came in last place, I believe or second to the last that year. I know Thea also ran. Yeah, the two of us were in last place. I remember some people said, "People will never vote for you because you're a woman and because you're from the United States."

Lovely.

So that was 2010 but here's the thing. As I'm pretty stubborn, so I kept running and losing but I kept inching up. So that year, I was in last place. The next year, I was in the middle of the losers. The year after that, I was first runner-up. So, I ran three years without winning. And then the next time, I won with more votes than anyone.

I don't know what that says about the community. It's hard to really read into that. And I did continue to hear the "No one's going to vote for you because you're from the United States." But anyway, people clearly got over that at some point. I don't know, I just felt like I had experience that could be useful.

I like to think I'm good at compromise and I'm creative. I always felt like a change in thinking needed to happen which probably if you asked me four years ago, I wouldn't have realized how difficult that sort of change in thinking would be. Just because I feel the Foundation needs to grow. I mean, I would like to see it have staff and I know that to some people that's a radical idea. But when people ask why HOT's effective, it's, you know, we have an operations manager where, for example, HOT was given a free booth at a Linux conference in California. I can say to her, "Can you organize the booth please" and know that it gets done. Or scholarship programs - having someone paid to make sure people get visas much easier. I think there's just a lot more you can do with administrative help.

So, what's your role at the foundation now?

Currently, I'm the chairperson of the board of directors. I was at-large board member the year before. My major goal over the past couple of months has been to facilitate having a face-to-face meeting of the board. I believe strongly that when people meet, it helps improve relationships. I'm hoping that's true.

You know, it's difficult because if you are a volunteer of an organization with a volunteer board and no staff, you have to make sure you don't take on too much or you have to find alternative ways of doing things. That's my role. I still want to improve diversity within OpenStreetMap.

There are initiatives that continue to go on but not really specifically through the Foundation. The foundation has never made a statement on diversity publicly that said, "this is what we believe" which is different from a lot of other software foundations. For example, the Python Software Foundation made a statement about not discriminating at least three or four years ago. Those are things that are expected in open communities now. Having a code of conduct and a diversity statement is considered like the least you can do and that doesn't fix anything. When people say that sexist things happen in conferences related to OpenStreetMap and then call them bloopers and do not respect the fact that women and minorities have rights as well, we're never going to be a free map of the entire world. If the goal is a free map of the entire world, anyone should be able to map and not stop anyone else, I think.

But is anyone actually being stopped from signing up to the website and mapping?

No, but you can imagine. A big part of OpenStreetMap is not just mapping at home without talking to anyone else though. And it's not unheard of to start mapping on the website and you're new and doing something wrong and get a nasty message from someone - that discourages people.

But I don't think that's limited to minorities. I think what Rich White Man does is like me getting nasty messages on OSM all the time.

That's totally true but then when the community meets in person or if you use your real name to communicate with people, there can be aggressive discrimination or purposefully treating someone badly. But there's a lot of micro-aggression-type behavior that occurs as well. If you say something like—someone once said something like "You map like a girl" which is like a very typical way of being sexist. It's saying you do something like a girl like that's bad. And the thing is, honestly, that's discouraging. Making it feel like a boy's club doesn't encourage new people to join.

I'm just curious as an aside, did someone say that to you?

Someone said that on one of the mailing lists but not to me directly.

I was going to ask, "Why don't you just slap them?" I mean, it's like—

Oh no, what I mean is if you were to look at the way people talk on the mailing list, there's instances of things that could make some groups believe they don't belong.

And that's by choice of language. And I think there's things that can be done. For example, Kathleen Danielson, one of the other board members of the OpenStreetMap Foundation, she's on the board of advisors at the Ada Initiative which their entire mission is to help facilitate women inclusion in open culture and open source. And there are different projects for different levels of—either try or don't try to make things inclusive. I don't think, with OpenStreetMap, we've ever made a really specific effort at being inclusive. And, you know, there's community cultures. Sometimes, I think it's gotten better but them I'm not so sure, you know? It's hard to measure.

Do you remember, there was like a conference a year and a half or two years ago where that guy made some stupid remark about a dongle. And then the girl in front, at this conference talk, turned around and took a photo of him. There was an article like two days ago or something in the New York Times or something about it. And in that circumstance, first of all, that guy lost his job and became a hate object. And then there was a massive backlash and the woman who took his photo. She lost her job, got fired. There was a massive backlash against her. And then as an aside, they had the other story about that lady who made a stupid comment about AIDS while getting on a flight to South Africa. And then she only found out right at the end.

What I wanted to say is if you look at—I attend a lot of conferences. And there's conferences where it's very clear, what's acceptable behavior and if something unacceptable happens what you should do, like who do you contact? Who's going to help you? And so like in that case, if that information was really clear. For example, I go to Linux Conf Australia every year. I went this year. I keynoted last year. And it's a conference where it's very clear, in simple things like you have the safety numbers. You know who to contact from the conference if you have a problem and it outlines what's acceptable behavior.

And so, in that scenario, if she felt like it was inappropriate and she's uncomfortable, having someone you can contact would probably have been a better outcome. But then, what did happen? I think both of those people would probably agree because both of them lost their jobs and, as you said, became hate objects. So I think, because there's much worse things that happen at conferences. People get raped. People get groped. There are terrible things that happen at conferences. You know, there's often drinking and things like that happen. And yes, some of it is a mirror of unfortunately things like that happening in society, but there's also a responsibility to at least make sure acceptable behavior is clear and then if something happens, how you can get help. I think that's sort of, to me, a minimum requirement.

Yup, the reason I'm asking these questions is because I suspect a lot of people don't really understand this. They think it's something that

doesn't really apply to them. I know such is just a part of the culture so it's like a fish swimming in water. It's like this is just how it is. I suspect that if anyone was to disagree with this kind of stuff, it will be things like "well, this is just the way it is or it's just reflection of society" Or it will be criticisms of things like misunderstanding the difference between equality of opportunity and everyone being equal like we all have to make the same money and have the same opinions. And often when people talk about equality, then all these things get conflated.

Yeah, I mean, there's this picture of these guys looking over a fence and they're all of differing heights. And equality is all of them are standing on a box to look over that fence that's the same height. And so the shortest guy can't see over the fence because he's too short. But then justice is having different size step stools or boxes so everyone can see over the fence in this scenario. And that to me is what the difference is between equality and justice.

So like when there are scholarship programs to attend a conference, if your employer's paying or if you can afford to go, then don't apply for a scholarship, but if people need that help, it's great to be able to provide that. A lot of people get defensive when you talk about these issues because they think, "Oh, you're trying to come after them" when really, you're just trying to make it so everyone can participate. The idea isn't to go seek people out. It's not meant to be punishment. It's meant to make it so that everyone can participate.

Oh well, I mean, I've been a woman in technology for 15 years now. Actually, probably more than that. I started programming when I was 10 and did computer science at the beginning of university and it's difficult. I think, in the OpenStreetMap community, some people view diversity as a North American issue. I've been told "Ah, we don't have sexism, we're British." I'm like, "Well, have you talked to any women in Britain?" I mean, I don't know, there's a lot of different cultural aspects in play as well.

What should we have talked about that we haven't talked about?

Okay. I think one thing to talk about that I think is important is, more than OpenStreetMap itself, it's also created these relationships that go beyond OpenStreetMap. I don't know how I call this a question but, you know? For example, Thea, like I'm the godmother to her daughter but we would have never met without OpenStreetMap and we don't do OpenStreetMap together anymore but we're quite close friends. I don't know if that's so much a question but there's a lot of relationships that have fundamentally changed where people have been dated or gotten married or in your case had children. I think that's interesting, you know?

I was at an OpenStreetMap conference, like it was a SotM Italy or something in the Dolomites. We went to a pizza restaurant afterwards

in this beautiful little Italian town. And as we were sitting down, this guy comes up to me and there's a lady with him and they just give me a baby and they said, "Please hold baby" and then they took a photo of me. I'm like "Okay, that's great. Why do you want a photo of me with your baby?" Like, "well, this baby was conceived at the first Italian mapping party. So without you, we would not have this baby." Like, oh, okay."

Yeah. I mean, it's the community. I mean, you know, a lot of people get together once a year and see each other. Some even more often, you know?

Steve Coast

Bio

Steve was born in England and founded OpenStreetMap while an undergraduate. Steve moved to the United States after founding a VC-backed startup, living in San Francisco, Seattle and Denver.

Favorite Map?

> *"The map of London GPS traces I produced with Tom Carden"*

Interview

Why don't you tell me a little bit about how you grew up?

I grew up in the southeast of England, in a tiny little town called Walderslade, in the Medway Town southeast of London. I had a very mixed childhood. My parents divorced a number of times from different people. I went to a lot of different schools. I had to live with my grandparents for a while, and my aunt and uncle. And then after a while, from about 14 to 24, I lived in Central London.

Were you a good student?

I was probably one of the worst students you've ever seen. Let me see, I think I got excluded from one school and expelled from one or two others. One of the schools, I just decided I wasn't going to go to—what in the U.S. I guess you call physical education, P.E.—because it was just a forum for beating the shit out of each other so I just decided not to go anymore and I got excluded from that school.

And then in my primary school, I must have been eight or nine and I was lighting matches or something. I got into a lot of trouble for that. Yeah, I didn't really pay attention. I was consistently told I was stupid for a long, long time so I believed it. Up until I discovered the library around I think I was then 14 or 15.

What were you reading about in the library? Was it computers from the beginning?

No. Everything you could imagine, things like old copies of *Scientific American*. I remember that one particular library had every issue of *Scientific American* going back to the dawn of time and you could request them out of archives. I remember a cover story about this new thing called the laser. This must've been in the '60s or '70s. A lot of old things like that. A lot of biographies, because you can learn an awful lot from biographies. You can sort of shortcut having to learn all those life lessons yourself by reading lots of biographies.

Any names in particular you remember? Edison or Tesla?

Tesla, a long time ago. I don't remember reading anything about Edison. Feynman would be one. I read all kinds of books. There was a good book on DeLorean, the guy who did the DeLorean car. That wasn't a biography but it was a history of DeLorean Motors. That was quite fun and you got to see these arcs of things, right? So, in his case, starting his company, figuring out how to make it work in Ireland, getting money from the government and then it all sort of blowing up in the scandalous way that it did.

I remember reading books on UFO's. I think *Scientific American* is how I first got into some of the computer stuff because there was this mathematical games column. I think it was Martin Gardner or Douglas Hofstadter. *Godel, Escher, Bach* was a pretty amazing book. And then lots of science fiction as well so Iain M. Banks, Isaac Asimov, Arthur C. Clarke. A lot of Terry Prachett. But I literally had this sort of transformation where I was just a stupid dropout and then discovered the library.

Yeah, I'm interested in that huge transformation. You're a bad kid. You're a so-called stupid kid, what drew you to science, math, and engineering?

Part of me wants to say that there's some sort of consistency in it and there's some sort of inherent rule-based system that makes sense whereas the rest of the world didn't make a lot of sense. Also it came very easily compared to other things.

Going into the college story.

I was going to a very, very bad school around this time. My school got shut down by the anti-terrorism police because somebody had brought a gun. That kind of thing happened all the time. And then, from when I was 16 to 18, I got to go to one of the best schools in the country. They took in bad people and spat out good people. After that, I went to UCL, University College London, which is like the first, second or third university in the country depending on how you count. I started off in computer science. And that was a lot of fun but it was difficult for me because I'd just come back from an internship with a company called Wolfram Research.

I had discovered this piece of software called Mathematica because in the library, there was a copy of the *Mathematica* book. I didn't have the software but just the book, and started reading through it.

One way to describe Mathematica is as a programming language, but it's actually much more than that because it can do a lot of things that are way beyond what you'd think of as a programming language so you can take a mathematical function and integrate it. You can plot things. That was all very interesting and I found their website and found they had an internship program in Champaign-Urbana, Illinois. And so, I wrote to them and then had a phone interview where they sort of figured out that I was, I think, smart enough to at least apply, but I didn't have any experience because I didn't actually have a copy of the software.

So, they sent me a copy of the software and then I built a bunch of test things with it just to show that I could do things with it. I sent that off to them and they said, "Okay, sure. Come over."

And then, in between the end of what you'd call high school in the U.S. and going to university, I flew over to Illinois on summer break and spent six weeks or so working on polyhedra. Polyhedra are things like cubes and dodecahedrons. Mathematica had libraries for how these things worked and interacted in 3D but they weren't particularly standardized at the time.

To get into that a little bit, so you could define a cube in a number of ways. You could say that the volume could be a certain thing and then all the different polyhedra have the same volumes. So you have a cube that has a certain volume of 1, then you have a pyramid with a volume of 1 but the volume dictates other things about the polyhedra. So, if your cube has a unit volume and then your pyramid has unit volume, it implies that the pyramid is a certain height and the sides were a certain length and so on. But if you go the other way and you say the length of one particular side is unit 1, it has implications for the volume and vice-versa. None of this was standardized and some of the things were symbolic. They had a polyhedral library that wasn't very uniform. I spent a bunch of time making it much more uniform and then writing programs that would, for example, take a polyhedra like a dodecahedron—which is made up of a number of pentagons—and unwrap it into what's called its net. So, if you take a piece of paper and cut out pentagon in a chain and then you could wrap them up into a dodecahedron, you could automatically open one to the other and you don't have to write software to do that. So, I was doing this sort of fairly hard, interesting mathematical/computer science stuff with some fairly cutting edge technology.

Then I went back to England to go into university and the first thing they taught was "Hey, this is a computer. You turn it on by pressing this button. We're going to learn about Java and this is an object in Java." And so, university quickly became very boring, but luckily I had an exceptionally good tutor, Jon Crowcroft. Now he's a professor at Cambridge, at the computer laboratory there. Through him, I discovered all kinds of other interesting things. Jon had high contact with students, especially smart students. And so, we used to go to a pub called Jeremy Bentham in London for beers every other day. Through that, I got to hang out with lots of PhD students and professors instead of undergraduates.

Well, before we leave Wolfram completely, do you have any colorful stories? I mean, I got an idea of the interesting kinds of problems you were working on and how maybe that experience changed your mind. It seemed to have kind of catapulted you ahead of your undergraduate class in computer science. Did you have any interaction with Wolfram or anyone there? Or do you just have a reflection of what working with polyhedral nets did to your brain? What about the emotional stuff too, like having fun—all of that?

Yeah, I was super young and naïve. I sort of got there, my first trip ever to the U.S., on my own. Wolfram Research had this excellent internship program. They put all their interns up in some apartments that were maybe a 20-minute or 30-minute walk from their offices in Champaign, Illinois. And then I got to live with these other guys and somehow found this second-hand bike. I used to bike around Champaign.

There's also some stupid stuff. I'd had too many Dr. Peppers at one point and I gave myself heart palpitations and I had to go the ER because I sat there just downing Dr. Peppers while writing software. And there was one time when I remember waking up in the office at like 3 or 4 o'clock in the morning and I'd fallen asleep on the keyboard and had keyboard impression marks on my face. But it was a very fun experience. I met Stephen Wolfram, I think, only once in the corridor for a very brief chat. He was just wondering "what are you doing here?"

I spent most of my time working with Eric Weisstein who wrote an encyclopedia of mathematics (Math World) which was great and very detailed and very interesting but the unfortunate downside was that Wikipedia was around the corner and knowledge—especially this kind of specialized structured knowledge was going to become very accessible, very soon. My memory is that Weisstein was trying to transform his encyclopedia from print to online, and Wolfram Research was helping with that that. In fact, the polyhedral work I was doing was part of that but at the time the internet was still relatively new, and applets for interacting with the web were new. People wanted to be able to do things like have a cube on the web and be able to spin it and see what it looks like from different angles but that was sort of brand new technology. We were building that into Math World at the time. But yeah, mostly I was just young and naïve and spending far too much time in front of the computer.

You watching as your friend was getting eaten alive by Wikipedia. Is there more to say about that because it obviously presages your future thinking about maps?

A little bit later, while I was in the physics department at UCL, I did an internship at a company called XRefer. At the time, they were taking multiple encyclopedias, linking them together, and then putting them on the web as a paid service for universities and libraries. For example, they would take a nature encyclopedia, which might have an article about beavers making dams, use software to link it to an article about hydroelectric dams from a science encyclopedia. And so you could explore knowledge in a way that didn't exist at the time using proprietary content. And then libraries could subscribe to this as a service for their patrons. It was doing pretty well but then Wikipedia came along and sort of nuked them from orbit.

Okay, so you're back at UCL and you're drinking a lot, but with the right people, and sooner or later you get to an idea that you're going to do OpenStreetMap. Is there anything to say about your time at university before we get to that story?

Computer science was kind of interesting. I was just flailing around. I didn't really know what I wanted to do and I certainly was not a good student in terms of doing what you're told, showing up to lectures. I was learning a lot and doing interesting things just not in the traditional way. My feeling was that if everyone else was doing it, it must be bad.

And you had this thought at the time? You just hated what other people were doing because they were other people? Or was this a strategy?

I think it was a little bit of both. I mean, certainly some of this is 20/20 hindsight. But also, I always wanted everything to be sort of applicable, you know? And a lot of what you spend your time doing at university you're never going to use in any context. Like in computer science, a lot of the theory behind object-oriented software or something, it's interesting but am I ever going to use this to make something real that's going to impact anybody? Probably not, so why bother going to this class? Which is interesting because when I failed computer science I transferred over to the physics department, which is even less relevant to reality.

This is interesting. Why did you do that? I mean, I find both of these interesting because physics—perhaps it's useless yes—but it seems to train people to think very well. So why did you go to physics?

I needed something to do after failing out of computer science. It was one of the most interesting options, and I did pretty well in it.

Did you say failing out of computer science?

Yeah. I mean, you never really officially fail. I just did my first year, then took my first year again and then didn't want to continue. So then, I just transferred over to physics. That's my memory. I spent six or seven years in university not getting a degree but still learning lots and meeting lots of interesting people and working—you know, I worked at two different PhD research labs at the university. I was the system administrator for two different groups. Also, I spent a summer at the computer laboratory at Cambridge, among things. But the actual core academia stuff just never really interested me very much. I was also typically just way ahead of people.

Physics was fun and there was kind of interesting stuff going on. I did a lot of computery things in physics. I remember the astrophysics department had its own little cluster of computers and I managed to get an administrator account to those and jury-rigged them into—back then there was this brand new thing about networking cheap computers together to make a supercomputer, so-called Beowulf clusters. I managed to do some interesting things with those. And while there, also, I worked at the Center for Advanced Spatial Analysis. I was doing things, making maps of the internet so—

Which was quite new at that time?

Oh, it was brand new. Some internet maps had been produced I think at Bell Labs. And they were kind of interesting and pretty but they had some fundamental flaws so I set out to improve them at this PhD research lab where I worked. That was kind of interesting and I produced these 3D movies where you could fly around the internet but the computational power required back then meant that you had to harness 20 to 30 computers at a time to do it. I sort of borrowed some of the physics department's computers for this.

When you're mapping the internet, you have a graph of the links between computers. What we did then, and still do today, is a spring-based force layout diagram. So, you pretend each link between these computers is a little spring. And then you pretend all of the computers are electrostatically repulsive so they push each other away. And that's what leads to those spider-like graphs of knowledge and the links between them that you see all over the place today.

And it's from doing that I ended up working at Xrefer. They found my work and asked me to come to talk to them. They had an applet that was doing the same thing. So, that example I gave you earlier about the links between a beaver and a hydroelectric dam. They had user interfaces with graphs that used something very similar to what I had done. I figured out that they had made some engineering choices which could be faster when made in a different way. I ended up working there for summer and then later full-time.

So, the internet was a different place at that time. Why don't you describe how you were essentially one of the pioneers, exploring this brand new thing?

The advantages that I and people my age at that time had was we didn't have any preconceptions. We didn't have to fit the internet into some other model like the yellow pages or the fax machine or whatever people wanted to use as a mental model for this information sharing, discovery and interaction. It was all fairly natural as a new thing. It's like watching a kid today just expect that everything like a TV is a touch screen that back then the internet, being a new resource to explore, you didn't come with any preconceptions about how it should work or what should be done. Back then, all kinds of things were being tried to figure out what was going to work for people. It came down to the simple stuff, basically a browser, web pages, and e-mail. This stuff has been around for a long time but people forget there were other technologies that were being tried.

There was a lot being done with 3D and VRML in the mid- to late 90's when VR was going to be the next big thing. There was Gopher. So it wasn't all just http. Today people either think Facebook is the internet or they think the web is the internet but it was much rawer back then. In fact, my computer science tutor had a copy of the landmark multi-volume book called *TCP/IP Illustrated*, which explains how the base protocol of the internet works in three volumes. It's pretty well-written. You start with this base knowledge that you just don't get today. It's a little bit like the difference between being a carpenter, and just buying a table at Ikea. So, you're going to have a different understanding about how things are put together and how things work.

This is another interesting thing. I think it was pretty much the same day, I got two letters. One was "Thank you. You've been at university too long. You're leaving now." And the other letter was "You've won the third-year prize for a project using computers." I don't know, I forgot what it was but I got 500 pounds or 1,000 pounds.

The project being the mapping project?

No. The mapping was just summer work that I did by myself for fun. The project was using Wolfram Automata to model gas flow over objects. So, it turns out you can set up cellular automata to make it look like a 2-dimensional cross-section of a gas and then you can blow that gas over something. For example, a wing and you can get lift out of it. So, as virtual air blows over objects you can get little cyclones behind it, counter-rotating cyclones. You get one flowing one way, then one flowing the other way. I managed to build a simulation of these in Java which is incredibly advanced for third year physics but if you have been reading cellular automata books for ten years then it makes sense.

And your interest in cellular automata came out of your *Scientific American* mathematical games interest?

To a degree, because actually a lot of that stuff was pretty old. Wolfram did it in the 80's, right? And computers have gotten a lot more powerful since then. The stuff you see today with machine learning and artificial intelligence, a lot of it was basically figured out in the '70s, but now you have the computational resources to do it.

It's like Charles Babbage's brass differential machines. It makes sense on paper. Unfortunately, you just couldn't machine brass to the specifications that you needed. And today, in retrospect, it all makes sense. You just have to wait for technology to catch up.

Okay, so you're failing out of school and you're doing all these interesting projects. And then, one day, you come up with the OSM project. Why don't you give us that, as deeply as you can remember it and hopefully in chronological order starting from the very beginning.

Sure. I mean, at the time, Wikipedia was taking off, and accurate GPS, which used to be restricted to the military, became more available to the consumer, but the handheld units tended to be pretty expensive. I got my hands on what we used to call a GPS puck. It looked like a hockey puck and it cost $100 or $200 back then. And you plugged it into a USB port on your computer. It was kind of crappy and took a long time to find a location but it worked.

So working in all these departments at university, I managed to scrounge an old laptop. At the time you could rip the CD drive out of a laptop and put another battery in there and then you could attach a GPS. You put the laptop in your backpack and you could go walking or cycling around and collect GPS traces. And so, I'd collect these traces, trying to build maps by myself. So I would bike around Cambridge, because I think I was at the computer laboratory for awhile there, and in London and I'd be creating these GPS points. It sort-of looked like Hansel and Gretel dropping bread crumbs.

From there, I figured out that I needed to make an editor to turn these GPS points into roads, footpaths and so on. And if I was going to make an editor for me, I might as well make it for everyone else, and put this on the web. It just seemed kind of obvious because Wikipedia had done this in the realm of text. They were making articles and then allowing people to edit them. I thought, then why can't you do this for maps, in a very similar way? And so, today we would call what I did iterative. But back then, it was just what you did if you had no money. It was putting things out on the internet, trying to get people involved, and seeing if it worked.

But it's worth mentioning, I think, that when I started OpenStreetMap, I started a bunch of other things with the expectation that just one of them would succeed, because you have this tradeoff that maybe one in ten new ideas tends to succeed. You have this sort of Pareto Principle, 80/20-, 90/10-tradeoff. OSM was just one of the things that I tried.

Another thing I built was an open source physics text book, because I was employed at some point by the physics department to turn some lecture notes into pdf's using some typesetting software called LaTeX. I was producing this stuff and I thought, "Well, why don't we just build all of this into a community-edited document so you could get a textbook out of it?" The difficulty there is that creating narrative is difficult. When you have collaborative projects, the narrative is pretty hard to build, which is why Wikipedia is so good because they don't have to build a narrative in those factual articles. So yeah, OpenStreetMap was extremely crude in the beginning. It still is in many ways. It went through a lot of iterations to get the technology basically right.

Why don't you just give me the actual technology? I mean, I understand when you say, "Oh, we built an editor," but what does that really mean? Is that like a Java Script program? I mean, let's get a little more granular on what actually you built.

Yeah. Java Script was still mostly in the future. At the time, it was a Java applet and it was very, very clunky and it was not user-friendly. It was on the internet but it was a nightmare of an editing experience. That was the first version.

Yeah, right now, I can just throw up a Google Doc and invite you and then a community to edit it. But how does the community edit text, much less map data? Is there a heroic level of programming? Or is it just something that you can throw together in an afternoon? I have no idea.

Well, I think, it was a little bit of both. It was heroic in its naïveté because there was a general assumption that it would never work in the same way.

But why wouldn't it work, because you couldn't program it or no one would be interested in this? Or you'll never get enough people to get critical mass? I mean, just take me through the pros and cons.

All of those things. An analogy to Wikipedia is worthwhile in that for many of the same reasons people thought Wikipedia was a retarded project that would never work. First of all, why would anyone volunteer their time to go do this when they've got to do things like pay their mortgage? And even if they wanted to, how would you create a software infrastructure that would work especially without a profit motive? And then how would you make this thing sustainable? And its sort of magic that it occurred.

And so, you had this entrenched industry that had been making maps in a very logistical, top-down way for a very long time. The only way you make a map by hiring hundreds of people who have to be highly-trained and use very specialized software. But these things tend to take on a life of their own and get detached from reality. Originally, you didn't need to have a license, and you didn't need to have training to make a map. But eventually training came into it, standards came into it, and people tried to build a system of knowledge around this, but the underlying reality is changing at the same time. So, as in the case of Wikipedia, the model had been Britannica where they said "Well, we have to hire experts from universities around the world to write articles. And then we need editors. And then we need a publishing firm and then a distribution network to sell these things."

Britannica evolved and evolved until the point where they didn't have connections with reality anymore because reality has evolved as well. They built in these assumptions and then the internet happened. And so, the same thing happened to map surveying. All of these assumptions that the entire industry was reliant upon no longer applied. For example, you needed specialized equipment to survey with range finders and triangulation and so on, and GPS made that obsolete for most surveying. And then you needed specialized software. Well, actually you just need some kid at the university to spend a couple of weeks writing something. And you need a strong ontology, right? So, a freeway is type 1., a road is type 2, a footpath is type 3. Where, actually no, you don't need those things anymore. The world is too complicated to put into an ontology anyway.

Then why don't you need that one—that's not as clear as the other two?

First of all, it's a barrier to entry, right? Let's say that you make some sort of surveying ontology schema that works in one place, it's inherently not going to work in another. An ontology made for the United States is not going to work in Pakistan where the road classifications are different.

An autobahn is not the same thing as a motorway, and it's not the same thing as a freeway. So, people tend to make country-specific ontologies, which is great if you're a national mapping agency, but if you're attempting to map the world, it doesn't work very well. You could either sit around and try and build an ontology that works, that covers the entire world, but that's really hard and expensive.

Or you can just throw away the idea of ontology?

Right. People, when they come across something that's already entrenched like the inherent need for an ontology in mapping, it just becomes part of their reality and it's very hard to break away from that. There are all these dependencies within the industry and these assumptions and it's very hard for them to break out of their mental models and accept the idea that you wouldn't need an ontology.

So this is what people were telling you, "It all looks great Steve but you forgot one thing like you don't have an ontology. It's just going to be garbage in, garbage out." Was that what you got? Or did you even know you needed an ontology before making this radical decision to throw it away? Was it just simple naïveté that you just stumbled into this, a surprising rejection of the mapping rule that you need an ontology when actually you don't really need an ontology any more than you need a newspaper to get your news?

Right, a lot of it was naïveté but then a lot of it was also just pigheadedness. Because pretty much everyone I showed it to was pretty universally negative, saying it would fail for some key reasons, either it didn't have an ontology, or it would never be accurate enough, or you would never get enough volunteers. But whenever you drilled down into it with people, they just had some stupid reason for thinking that. And there wasn't any bottom up argument for that being the case. It was pretty much "Well, we've not done it before, therefore, it's not going to work."

And so, you were showing it to mapping people or professors or who?

Everyone.

Because you were really proud of it, it sounds like. Why else would you be showing it to everyone?

It wasn't quite so much pride. It was just like, "This thing is obviously going to work."

Well, how did you know it was obvious? I mean, we stopped the story and veered off into theory when you were just walking around London with a laptop in your back and maybe you did three weeks of programming for a group editor. I mean, what happened?

A lot of these ideas were either extrapolating from the technology, or extrapolating from Wikipedia. GPS's were going to get smaller and cheaper. Computers were going to get faster and cheaper. And then Wikipedia had clearly proven this model for text, and Linux had proven this model for software. So why wouldn't this apply to mapping data?

And who's we? Now, you said we, like, I thought this was sort of—I have you in my head as a one-man band, but—

A lot of those things came a little bit later. So, first, there's Steve running around with a laptop, showing this to people but there were barriers. Today, we have iPhones with GPS and practically everyone has one of some type in the western world. But back then, the audience of people who'd be interested in this is one thing, and then finding people who also had a GPS would be another, and then finding people who were interested, had a GPS and were willing to survey was even smaller.

I used to give a lot of talks at user groups—Linux-user groups were the big ones because Linux, at the time, was fairly big and there would be user groups where people would do talks about new things that they were doing with Linux. So, "Hey, I bought this tool or I saw this piece of software, I got Linux running on my calculator." And there was a lot of crossover with that community because they were already comfortable with the core ideas of community and open source. Also these people were comfortable with spending their spare time doing these kinds of things. So, it was much easier to convince the Linux people to buy a GPS and play with OpenStreetMap and build some software, than it was to go talk to mapping people who did this as a day job, and didn't want to spend their spare time mapping as well. So, Linux users were a far more receptive community.

So, it's very clever. Just to be clear, were you using Linux to build OpenStreetMap in some significant way?

Yeah. In the beginning, Linux was pretty much exclusively the base technology everything was built on. It's not quite so true today because there are all kinds of things that don't need Linux, but back then it's what was used. So, once I convinced some people to start using it, which was fairly early on, then we tried to make it scale as a project. And that really involved opening the door and letting people run with it and create things that were very simple and platform-like, to free people up build whatever they wanted to on top. This meant that OpenStreetMap's API is very simple, very open-ended. It went through some iterations. In the beginning, it was all Java with XML-RPC. Later, it became Ruby, then Ruby on Rails and REST. REST was also a brand new technology at the time. But it was just primarily getting out of the way, and creating something that was really easy for people to take and extend as a route to building something larger. Because if I tried to do everything myself, it clearly just wouldn't work.

Now, in the United States of course, the people own the maps that are created by the federal government agencies, I understand that's not true in England where you're from?

Exactly. That was also one of the drivers. I had this laptop, I had a GPS and I wanted to see my location. And you couldn't do that because the map data wasn't available. You could infringe on copyright and download some pictures of maps and you could see yourself on top of the picture but the computer couldn't do anything with that picture. It couldn't tell you the name of the street you were on. It couldn't tell you to turn left or right or anything particularly useful, and you were infringing copyright by downloading those images. So, it was kind of a necessity that you had to go out and create your own vector data in order to do anything interesting at all.

Oh, I see. So you got this cool new toy, your first GPS that that was "de-fuzzed" courtesy of the American government, and then you realized you didn't have a map to run it on to make sense of the coordinates that you're generating?

Right. So, in Britain there's a national mapping agency, the Ordnance Survey, and they have a very long history of making very detailed, very good maps. These maps were just largely unavailable to people like me. If you were a very large academic institution you might be able to get access to data, or if you were a car company that needed a map in your car, then you could pay. And even if you paid, you'd get the map data under some fairly difficult license. It would be like asking Britannica, "Hey, can I have a copy of your encyclopedia to play with?" Like, "Well, no. But you can pay $1,000 for the 28-volume set." It was the same with map data. So you'd have this GPS which you could show where you were anywhere on the planet. There was just no map data to give you a frame of reference or to allow a computer to do anything interesting with it, which necessitated a map. And then if you're going to create a map, well looking around, what are the analogous things that are working? And Wikipedia was the closest thing.

Were you the only guy with this idea?

I was not the only guy with the basic idea but I was the only guy to take it through to its logical conclusion. The projects that existed, they tended to be country-specific. "Hey, let's make a map of Israel" or "Hey, let's make a map of England" so there was that geography-specificity to it, but there was also a data type. So, it'd be like, "Hey, let's make walking maps of England" or "Hey, let's make road maps of Israel." There were those two restrictions that OpenStreetMap did away with because Open Street Map focused on the whole world for every type of data set.

In the age of the internet, all of those distinctions don't really make sense any more. If you have a national government funding something, then "Sure, let's make a map of Spain." But if it's something on the internet where anyone with a GPS can contribute, it makes no sense to restrict yourself geographically or by data set, you might as well just make it open.

And then, the mapping projects of the national governments tended to be technology-led instead of people-led. Let me try and give you an example. If you're going to build a map database based on technology, then you would make some very complicated stack and you would have a lot of formality like roads can only connect in certain ways. You would restrict the types of data that people can put in and you would try and use the existing models of creation to do this.

It would be like trying to open Britannica's editorial process and just letting volunteers run the editorial process as opposed to just letting anyone on the internet edit some text as they see fit. So, if you were to open up Britannica, you would say, "Hey, I need someone to volunteer to be the editor of fish." And then they would try to find volunteers to write complete articles about sea bass or something instead of just allowing all of that stuff to be bottom up and self-organized and let people work on whatever they want. A lot of the projects at the time, they took that approach. They say, "Hey, let's look at the existing models of how maps are built today and let's try and somehow merge those into the internet." And what that means is "Hey, I need a volunteer to go map North London and you've got to use these data types, and you can only use these tools." Whereas, OpenStreetMap did away with all of that, and just said, "Hey, go map however you want. We don't care. And map wherever you want, we don't care." So those barriers to entry went away.

So, famously, when Wikipedia was founded there was some sort of review before an article was published and in the beginning it didn't get much traction. And, almost in desperation, they said, "Oh, let's just open it up. We'll just publish it first and fix it later." Were you aware of that history? Did that affect your thinking?

I certainly wasn't aware at the time that Wikipedia used to be closed and they tried to pay people to create articles. No. I don't think, at the time, I was aware of that learning process. I just sort of took it as a given that that the simpler thing tended to be there from the beginning.

Well, the national map projects that involved volunteers had similar editorial processes. You would take a GPS trace. You would turn it into some sort of PDF or PostScript file. Then you would e-mail it to someone who would look at it, would integrate it into their PDF and then send you an even bigger PDF. I saw all of these things as just simple barriers. There should be as few barriers between you and the map as possible because any barriers you introduce are just going to reduce the number of volunteers you attract. The starting point of the national map projects is that you don't need many volunteers. You just need a few specially trained people, instead of lots of untrained people. And so, there's implications of this way of thinking that influenced the way those projects were implemented. OpenStreetMap took a very different tack.

And you just came to this people-centric approach, how? I mean, you're not a people-centric person really. You're a science nerd, maybe even a loner—spending your free time running around with computers running in your backpack. Was it just somehow intuitively obvious to you as someone who's been soaking in the internet since its very beginning? Or was there some lesson you learned from an earlier experience in your life?

I think it fell out of the point of view that you want lots of volunteers to do this and the only way you're going to convince people to voluntarily go spend their time on this fairly esoteric activity is by making the barriers to entry as low as possible. And if you're going to do that, then that drives your technology decisions, whereas the national map people tended to be the other way around. They were, "I'm going to make a bunch of technical decisions and hope that people conform to that and show up."

Okay. So, tell me about that the first glimmers you had of "Hey, this is going to work. This is going to scale."

I always thought it would work because the model had clearly been proven by Wikipeidia, but I wasn't sure if it was going to be OpenStreetMap. And so, the question was just, "Who was going to figure out right model for mapping?" But with OpenStreetMap, specifically, I'd invented these things called mapping parties. I remember standing in an office with Tom Carden, and we were looking at a map of the U.K. and had noticed that there was this island to the south of the U.K. called the Isle of Wight. We thought we could go try and map the whole island. So, the project was starting to take off. There were people. There was a mailing list. People were talking about mapping. They were mapping individually but we hadn't done anything to bring them together.

And so, the Isle of Wight was the first attempt to do a couple of things. The first was to bring all the key people in OpenStreetMap together. The second was to try and map a finite area in a finite amount of time. The Isle of Wight was kind of perfect because the number of people involved, about 40 people—could reasonably map it. If you have an island, it's a defined area.

So, we had a weekend mapping party where everyone came together. It was a very boozy event. People went out in the morning to map, and then went to a pub for lunch. Then they'd map in the afternoon, and go back to the pub. It was a very social activity because you were creating social proof that other people were as crazy as you. And you were also creating an informal community/structure/hierarchy for people to be a part of.

It wasn't just a little convention. It was like a breakthrough.

Right. There was a product at the end of it that we could all point to and that didn't go away. It wasn't ephemeral. It's not like we all got together and had a great dinner. There was an artifact at the end, so that was one important early lesson.

And then the other key lesson was that in the first few years of the project, to convince people and show them how OSM worked, we would sit at a computer with a flaky internet and I'd ask them to zoom the map into somewhere that they knew, because when you know somewhere you're far more likely to have something to say about it. When you zoom in, you typically find something that is wrong so you'd say, "Do you notice anything that's a mistake? Anything missing? Anything we could change?" And people would typically say, "Yeah. Hey, there's a road missing" or "That road doesn't actually go through there. There's a footpath." And then you'd hit edit and you'd show them how to map. And then you would submit those edits and then they would be live on the map. And today, that process, of making an edit and the map updating, is less than a minute. It's seconds. But back then, I mean, it could take days for the edit to appear, if at all.

Because I've done a lot of these sessions and people would typically pick places where they'd lived or gone to school; New York, London. It would all be fairly predictable. I remember someone saying, "Hey, yeah, let's go to Cuba because I just went there on vacation." And so, before even doing anything on the computer and zooming into Cuba, I prefaced it, I was like, "Hey, London's going to be great and New York's going to be great, but there's probably nothing in Cuba, and there's probably not very much aerial imagery, so don't be disappointed when we zoom in and we don't see anything." People extrapolate from that. So if we zoomed into Cuba and there was nothing, they would just assume that the whole world has nothing. Whereas, actually, there's lots of activity in Europe and the United States.

So then we zoomed into Havana and I'm expecting to see nothing. And the thing was complete. All the roads were done. The hospitals were there. There were one-way streets. There were even buildings. And that was the point where I kind of stopped being surprised at the growth and future of the project, because there were always these little surprising moments. Like, "Hey, someone figured out how to map bridges." And then "Oh, Hey someone figured out how to map benches or lighthouses." All these strange things you can map around the world. And when I showed this person the map of Cuba and it was already done, that was the point when I could no longer be surprised. And also, the future of the project was fairly well assured at this point because if places that you wouldn't even expect to be mapped are done then what is left, you know?

I mean, it's just a question of time to create the depth and the breadth that you need—the breadth being the globe. So not just the U.K. but Cuba. And then the depth being "Hey, we don't just have freeways, we also have footpaths and benches and parks," and that level of depth. And everything from here on out is just going to take time. But there's no sort of fundamental constraints that are going to mean "We're not going to get Cuba, or we're not going to get footpaths." It's just going to go on and on with enough people and enough time.

Let's go back to that Isle of Wight story. This was a very clever idea that you had because that's both the first convention and a big win and the first big moment. But socially how do you explain decisions that were more than just technical. Do you have any fun memories of that time? I mean, were you sleeping in tents? I mean, what was the scene?

Honestly, just a lot of drinking and far too much arguing.

About what?

Well, arguments are dual-sided. First of all, in an alcoholic environment, there are naturally arguments. But also, one, its very positive that, you argue out the details of the project. Like, should it be built this way or that way? Should we run things this way or that way? And you get to slug it out in pubs and discuss that stuff. And two these arguments also create community, because if we had nothing to argue about, it'd be pretty boring. So, having those arguments is both important and positive but it has this flip side that it alienates people, that people can feel hurt. And also, I was a lot younger at the time. When I first started this project I was 23 or 24. I think I was 23. And dealing with a lot of people two to three times my age had gotten to the point where sitting in pubs arguing wasn't necessarily the best use of their time, especially doing it with young naïve people who were trying to take over the world with some mapping project. So, I alienated a bunch of people.

At that point? On the Isle of Wight, you did this?

Well, certainly there and other places because the whole project revolved around pubs. Looking back, the simplest explanation, I think, is that there's an existing culture of drinking in England of people going to the pub together and all we're doing is latching onto that existing culture and just bending it to our needs, which was mapping. And people have done similar things in different places.

It was actually one of the problems in trying to make OpenStreetMap take off in the U.S. We were trying to apply the model that worked in Europe, of people getting drunk together and creating a society, but it doesn't really work in the United States because the same drinking culture just doesn't exist. In Europe, you can expect people to get home by biking or train. And in the U.S. everyone's driving so they're not going to drink because they don't want a DUI. And also, the types of mapping are different. In Europe, you can walk and bike a lot of places to map so there's a lot of walking and biking people. But in the U.S., biking around most suburban neighborhoods is suicidal, so the way that people map is very different.

Well, it sounds like the beginning of a real social structure came out in this Isle of Wight, I imagine. Do you want to name some names, who were the other important early people?

Well, there were all kinds of people in different roles. Richard Fairhurst ran one of those competing projects that became part of Open Street Map. A lot of key technology people like Tom Hughes, Matt Amos, and Grant Slater. Who else? There were people who helped run a conference called the State of the Map. There were a lot of early people involved in that. There was Martijn van Exel, Hank Hoff, Andy Robinson. The list of people just goes on and on. The project wouldn't exist without them.

Is there any more to say about you as a naïve 23 year old? I mean, running what's fast becoming a global project?

Well, the primary naïveté that I had was, as the project was growing into something important, there was a demand to build a structure that was bigger than just "Steve deciding everything" which was very reasonable. And so, we created a foundation. I mean, I made a number of naïve decisions right there about how the foundation was structured and built. I just had this assumption that everyone thinks the same way I do. So, my model for the foundation was predicated on other people giving up control, sort of stepping back.

As the project was growing, I gave away a lot of control. I used to do everything, like I wrote all the software. I started the conference with other people's help, of course. I gave all the talks at conferences and events. I managed the community. I ran the mail server—I just was doing all the things that went into the project. And I had this idea that "Hey, the only way that this is going to scale is by allowing other people to take over control and for them to bring in new ideas and to build the project like that." My expectation was that in giving away pieces of the project, that other people would do the same thing. You give them some piece of the project and they would similarly divide that up as the project grew, give other people control, and incorporate new ideas—even ideas that they weren't comfortable with, and help those people grow in the same way that I was helping them grow.

And that, for the most part, simply didn't happen. And so, because of that OpenStreetMap hit a wall. You wouldn't see it as an outsider and you wouldn't hit it unless you had high expectations. It led to a lot of the similar things that you see in Wikipedia. For example, Wikipedia pretty much is still the same website from five years ago or ten years ago. If you were to transport yourself back then, there's not a whole lot of difference between Wikipedia today and Wikipedia from five years ago. The same is true with OpenStreetMap. That was kind of painful and disappointing because there's a lot that we could be doing.

It would be like saying that you would never say, "Hey, I look at an iPhone and it's the same iPhone from five years ago, or I do a search on Google and it's the same search today as five years ago." No, there's some really massive fundamental differences in how those things work and what they've done. Google went and bought YouTube or Google bought out the Android operating system. Or Apple added touch ID or instant payment using RFID with your iPhone. These sorts of fundamental new ways of doing things and fundamental new features that you don't really see in any open source project. This isn't unique to OpenStreetMap. But you don't particularly see Wikipedia doing anything new. It's the same old stuff. And in OpenStreetMap, it's the same. That said, it's grown remarkably well. It's just that I'd love to find a way to make an open source project create new things. But the way it seems to play out is, someone goes and builds something new, creates some sort of community and infrastructure around it and then it's kind of just maintenance for a long, long time. And you see that in most open source projects. And then you just wait for something new to come along.

Maybe a reasonable analogy is Kuhn's *Structure of Scientific Revolutions*. There used to be an idea that science was this plodding, continual quest for the truth that slowly got better, but actually it looks much more like punctuated equilibrium. You get this long, long period of Newtonian physics and everyone thinks that we've solved the world with the exception of some edge cases, and then along comes relativity and kind of blows that up. It's the same way with open source. You get these large projects that just become kind of sclerotic and boring and do the same thing until something new comes and blows them up.

I want to pull you back to kind of the earliest mapping parties and the early phase. You described what happened, but was there what someone might call a power struggle going on or was it simply you stepping back?

Well, it's not defined as a power struggle because it is a volunteer project. There are these de facto ways that volunteer communities work, but there's not that formality that you would get if there were some sort of political battle within a large company. It doesn't work the same way because you're not really competing for resources that are free and voluntary anyway.

But aren't people competing for the status?

Well, not really because status is not a very tradable asset. It's more that you can't fire volunteers. There is never a clear delineation. There is never, Bob is VP of marketing or is not VP of marketing because in open source, there are just collections of people and there's just de facto ways that they're treated. And so, the structures and the power struggles tend to be very different. Once you're entrenched and you have something key, I mean, it's almost impossible to unseat you. What can you do? They're a volunteer. You either leave and start a new project which people do, or you just give up and submit.

There really aren't traditional kinds of power struggles. And don't forget that all of this typically occurs over fairly open channels. There are public mailing lists that this stuff happens on. It's not like office politics where I go down the hallway and I go into someone's office and we start bitching or plotting. There's a lot of it that's done very openly. And when it isn't done openly, other members in the community tend to get very confused and upset and start asking "what's going on?" In open projects there are power struggles but the methodology and the results are very, very different because even at the end of this, let's say you are successful and you wrest power away from someone, all that means is you've got more volunteer work on your hands, right? So do you really want to do this or not?

It is my impression that there's kind of an inner circle at Wikipedia that makes the important decisions. Was there not a kind of—like you said one person can't run a global organization, so I assumed kind of

some sort of inner circle came out of the Isle of Wight or around then? Is that right? And if so, it's important.

Yes, there's an inner circle but you've got to remember the scale-free nature of this. It's very analogous to what Jaron Lanier says about the concentration of wealth and power in the internet age. One guy, Mark Zuckerburg, can create this thing that has enormous horizontal value and concentrates a lot of the money and power in one place. In the same way, in OpenStreetMap, the core of it, the technology that enables all of this magic to happen, is incredibly simple and can be controlled by one person. You don't have the same group dynamics because of that concentration of volunteer power that in a normal project—

Today to make something with enormous horizontal power, such as Facebook, actually requires really very few people. And so, you don't have that same spread of volunteerism. And things can either succeed or fail based on a few people—very, very, few people by historic standards.

Do you have more stories about cultural barriers as OpenStreetMap grew to a global thing?

Well, there are a few things. I mean, one that's analogous is when the project was starting, everyone was using handheld GPS units that cost hundreds of pounds and you'd have to have a digital camera as well to take photos. And so, if we're building map data, what's the connection between OpenStreetMap and what you're capturing? Because ideally, we would have your GPS trace and we would have your photos as well, but we didn't have the resources to store all the photos. We couldn't become Flickr. And so, that became a cultural thing because people would take photos in different ways and it drove the way that people mapped and the way that they explained what they were mapping because this stuff has to be verifiable. So, people would map in different ways.

I had an expectation that one person would create the GPS traces, another person would turn it into roads and then another person would add the road names. Or at least, that was possible. But it never happened like that. It was much more vertical. One person would do all three. They would go out, collect the data, then go home and spend about the same amount of time editing it into the map. So if you went out and collected GPS trace for an hour, that simply meant you had to spend an hour on the computer turning that into a map.

In terms of different geographies, it was surprising how Germany took off so spectacularly compared to the rest of Europe. It was surprising that in places where mapping is illegal or there's supposedly not very much infrastructure, that mapping took off, for example, in Cuba. There's mapping in North Korea. There's mapping in China where it's illegal to map.

So, it's illegal the map in Cuba and China?

It's illegal to map in China, certainly without a license. And in Cuba, I don't know whether it's illegal or legal but they only got the first internet line recently. There's an assumption that they're driving around in 40-year-old cars, why on earth would they have GPS and laptops? Maybe that's just me being a naïve westerner. I was focused globally from the beginning of OpenStreetMap but I didn't really think very much about places outside of my cultural experience. So, my experience was Western Europe and North America.

Sure, you could go use OpenStreetMap in large parts of Africa or other places in the developing world, but the idea that they are able to leapfrog all of the bureaucracy was incredible. So, if you're in Zimbabwe, you don't need to follow the same path that France or England did "Hey, let's set up a national mapping agency. Let's hire hundreds of people." They could just shortcut that in the same way that they didn't need to install landline telephones for everybody. They could just jump to wireless. There are unexpected things that OpenStreet Map as a platform enabled. That was the whole point. It was to do things that you couldn't predict.

Was there more mapping in places where mapping was discouraged? More than, say, in places that are already covered by good commercial maps.

Yeah, I mean, you'd be surprised because the tradeoffs were different. Because if you were in some remote part of England, you could just go and buy a map. But if you were in some remote part of Africa, your only option was to collaborate with other people and build a map. So the incentives were actually different. In England people might have more money and more free time and GPS units to go and map somewhere, but you don't have the same inclination if maps simply don't exist, right?

So, you have the best maps of places where the worst maps or no maps existed?

In many cases, yes. There are people mapping slums in Africa. These slums officially don't exist and officially there were no maps because they don't exist. Clearly the best map was going to be OpenStreetMap over and above anything else because your nearby government had decided that you don't exist so there was nothing.

Any actual places that come to mind that don't exist except OpenStreetMap? That's fascinating. Or are there any other fascinating regional differences?

Well, there were these cultural differences. I might be wrong, but it seemed to be in Germany that the culture is much more "let's get together and decide everything first. And then once we all agree, then we'll all just go do it." I don't know what it is but a good example would be that in OpenStreetMap that tended to be the debate strategy, like we spend a bunch of time, think everything through, and then go do it. Whereas when I started the project and how I operated in the U.K. and other places was much more "just go do whatever you want and we'll figure it out as it goes along." And so, there were definitely clashes because of those different ways of looking at the same problem.

For example, ontologies. Let's decide how we categorize everything before we map it and then we'll go map it. Rather than let's just let anyone do whatever they want and let it become an emergent phenomenon because despite all of these things being open-ended and you can do whatever you want, there is severe pressure for you to conform to certain ways of doing things because just from a rendering perspective, if you don't map things using the same ontology and using the same system as everyone else, it's not going to get shown on the map.

There's a lot of map data in OpenStreetMap that just doesn't show up on the map for a couple of reasons. One, the map would be too crowded with stuff. Secondly, there are things that people map for themselves that others are not interested in, or that don't conform to the schema that everyone else is using. An example might be administrative boundaries. You may be used to certain country and state boundaries but there may be lots of other boundaries that exist that you might not necessarily want to or need to show on the map.

There may be historical things or cycle paths that don't show up on the map. There is loads of cycle path data on Open Street Map that isn't shown by default because it would overwhelm the map. But you can make a special rendering that just shows all the cycle-related information. And there's tons of information such as the type of cycle lane, is it on the road? Is it on the sidewalk? Is it one way cycling? Is it two-way cycling? A lot of those things are in in OpenStreetMap but don't get shown.

What about turning points in time, the historical breaks or moments that are important that affected your endeavor or your project?

There were a couple. When OpenStreetMap got aerial imagery, that was a game changer. You used to need GPS or a phone to walk around collecting data for OpenStreetMap. But if you have aerial imagery available, it means that anyone can just look at a picture and draw on top of it without having to leave their home or office. And that led to basically a massive step change in the availability and quality of the data that was available. That was done when I went to Microsoft. Microsoft Bing Maps gave OpenStreetMap a license to map over the aerial imagery.

So you convinced them?

I would love to be able to claim that, but it was a bigger decision than mine. It was a question of what could Microsoft do to help the project grow. It was something that was very convenient because it wasn't throwing money at a bonfire. It was using an existing asset and just changing the rules slightly to have this massive impact.

When was that?

About 2010, I think.

Were you ever kind of self-employed, full time as an OSM person?

Yeah, so I started the company, right? As OpenStreetMap was taking off, it was fairly obvious that we could provide services and support around the project to people.

And the analogy would be with RedHat?

Exactly. So, RedHat, at the time was relatively new, but they were providing exactly that kind of service and support around Linux. If you were Steve, the university student with a lot of spare time, then you could go download Linux off the internet and figure it all out in your spare weeks. But if you're United Airlines or something like that, you need someone to rely on to make sure all of this stuff works. But the savings from the free software could be negated because you still needed to spend a fairly substantial amount of time on integration work and if it crashed, figuring out why. And Red Hat was providing these services around Linux.

I thought, why couldn't you just provide these services for other open things like OpenStreetMap? Unlike Wikipedia, which is the place where you put all the data in but it's also the place where you read the article. With OpenStreetMap, it's slightly different. OpenStreetMap is where you put the data in but you might consume OpenStreetMap any number of ways. It might be on your phone using a third-party app. It might be you're looking at an OpenStreetMap printout on the side of a bus shelter. It could be all kinds of different kinds of ways that you're using it that are not core to the project itself. It was much more analogous to Linux. So why couldn't we provide those services?

Is there anything else we should cover about early days—anecdotes, fun stories or turning points?

I think the conference, maybe. After a couple of years, it was fairly clear that there were lots of people interested in this project and wouldn't it be good to have a conference to get everyone together. And so, I called it the State of the Map after the State of the Union, of course. It was held in Manchester, England, in 2007. Eighty people showed up, mostly from around the U.K., with a bunch of people from Germany. A fairly standard affair. My memory is it was over a weekend. We had a bunch of talks. Very low cost. You know, we got the venue pretty much for free. It was just one of those first times when we got everyone in a room together and they could put faces to names on the internet. The OpenStreetMap mailing list had always been pretty vitriolic. I remember meeting one particular guy who had told me a lot of stuff about how I should live my life and so on. And then when we met at a pub in Manchester, it all just kind of disappeared because it's much harder to do any of that stuff in person.

Well, anything foundational come out of that? A coordinating committee, or a new direction, or something like that?

To be honest, I don't think so. I don't remember anything. It was just the first of many. Now, we have State of the Maps and are thousands of people they cost hundreds of thousands of dollars.

You told me how OSM just didn't take off in the U.S. in the same way it did in Europe and specially England because of the kind of the drinking culture and I didn't ask. Well, how did it finally get traction in the U.S.? Because, clearly, the U.S. map is pretty good.

Yeah, I think part of the problem is just statistics and reporting. When you look for activity in Open Street Map in the U.S., it follows population density so it's cities right? But the densities are just so much higher in Europe. London is just this ginormous urban sprawl compared to Colorado where you've got Denver in the middle and then basically nothing else. So, I think that the first thing is just that there's massive differences in the sort of over the environment--the layout of the cities and the sparseness between the States.

Did it really take off in the U.S.? I mean, it's arguable that it didn't. I think it's also arguable that it did. It's arguable that nowhere really has caught up to Germany in their level of mapping. I'm not sure that's really necessary because there's this big tradeoff that if you're mapping the cities really well–well, that's where 90% of the people are anyway so you don't necessarily need to have the rest of it mapped brilliantly. And you see that with commercial map providers as well. Their maps of urban areas are much, much better than anywhere else outside another place in the U.S.

We've had conferences in the United States. Those have worked relatively well in bringing the people together. But again, it's a big place, right? The mapping parties thing is done on and off. People have weekly or monthly mapping parties. There are OpenStreetMap groups on meetup.com and things like that where they'll have regular events in different cities, different states. But it doesn't have the same cohesiveness. It might be better to talk about OpenStreetMap in California or New York or something. There's simply not the same level of cohesive community that there is in various parts of Europe.

Okay, tell me about this earthquake thing.

Well, the beauty of OSM was to let people make their own mistakes, let people build things on top. Let people go where they want to without restrictions. OpenStreetMap acted as a platform for people to go do things. One of those unexpected things was disaster response, disaster relief.

At the time of Hurricane Katrina different aid agencies would make their own maps. So, you have an aid agency trying to provide shelters and they'd have their own maps. An aid agency trying to provide water and they'd have their own maps and so on. Each one of these had their own little mapping departments and had their own areas of focus and types of data that they needed on the map.

There had been efforts to centralize those over time but they were resisted because charities, as it turns out, are really just businesses. They don't want to give up those resources. They rely on having those resources to get their funding. If they don't have those people, what are they going to do? You know, people resist change.

And so, the efficiencies weren't there. Everyone having their own mapping department didn't seem to make a lot of sense. One anecdote from Katrina is that Interstate 10 over Lake Ponchartrain got destroyed along with a railway bridge that ran parallel. The aid agencies were using Yahoo or Google Maps to route their supply trucks to get into the disaster area to provide relief to people. And the maps were obviously wrong because that bridge didn't exist anymore. And so, Mikel a colleague who's also interviewed in this book, phoned up and said, "Hey, that bridge is missing." But they couldn't do anything about it because Yahoo's upstream data supplier was the ultimate arbiter of that data. And so, it took a long, long time to get the map fixed. Whereas If they'd used OpenStreetMap they could have gone in and removed the bridge immediately. And so, it's all fun and games until you figure out that the trucks full of supplies sitting on the side of the highway because of bad map data leads to people dying or getting diphtheria.

And so, Katrina was a bit of a miss from an OpenStreetMap point of view. You can fast forward to other disasters like Haiti where OpenStreetMap was more prominent. Haiti's government mapping agency building had been destroyed by the earthquake, so they had no maps. They had no infrastructure for distributing relief. OpenStreetMap became central to the relief mission.

There were layers of people contributing. The aerial imagery/satellite imagery companies made georeferenced imagery available to OSM and others. Normally, they would charge a lot of money for those images. People who were remote took the imagery and started digitizing it because you don't have to be on the ground to digitize buildings and roads and camp sites and whatever other features you see in the imagery. And people on the ground could take that pre-digitized data and annotate it with names and other information. Via the combination of a partially functioning internet and people using pen and paper to note down information, you could get this data into OSM but then you could also get it back out. So, people doing search and rescue in collapsed buildings were able to take OSM data and put it straight onto their GPS devices. These don't look like phones. They're hardened, with low-CPU power and big GPS antennas. And they could put the GPS data on there and go from building to building and figure out "Has this building been searched or not? Where am I?" because they didn't have any of the traditional infrastructure to rely on.

There were feedback loops of people actually taking the data out and then using it in their traditional scenarios such as searching buildings for dead people. And then when they found mistakes, going back and being able to update the data. And that, as an outsider, appears to be a fairly large game changer in the way that we respond to these kinds of events globally. But response to disasters is highly variable. Katrina and Haiti were well publicized but many disasters just don't get any particular attention. From what I can see, the funding tends to follow that. And so, the amount of map data that you get tends to follow that too.

How about more on for-profit companies using OSM as a platform to build interesting or surprising things. Surprising either in what they build or surprising in how well they've done as an enterprise.

It's hard now because it's just so ubiquitous. I went to the ESRI User Conference a few weeks ago. The main thing that stuck out to me is how everyone was using ESRI tools but they were using an OpenStreetMap base map primarily because it is free.

I remember OpenStreetMap being used on bus shelters in Europe where public transit's a big deal. That's one of those perfect use cases because even early on, OpenStreetMap was good enough for printed paper maps, and the cost was zero.

Do you want to talk about the company you started with OpenStreetMap?

I had started a company with a couple of friends in London—this was the forerunner. It was a classic thing. The three of us were going to quit our jobs and start this company. We had a lot of ideas. And then I actually quit and those guys didn't. And so, I had to try and figure out how build revenue, basically by doing consulting work and projects here and there, and just having a magical belief that it was all going to work out.

And through that, I met our seed funder. Open Street Map was taking off. And so, this particular guy said, "Hey, why don't you do a company around this OpenStreetMap thing and I'll give you some seed money." It was fairly obvious that you could take OpenStreetMap and provide products and services around it in the same way that was done by Red Hat with Linux.

All of the services needed to integrate and use the data could be done in slightly different ways with OpenStreetMap because the map data isn't locked up, the barriers to entry were much lower. So, to get map data, you didn't need to spend months in a sales process to go buy it. You just downloaded it. And so, that would inevitably lead to new business models.

I also met my co-founder, Nick Black at UCL when I was doing system administration for the Center for Advanced Spatial Analysis. I think Nick was taking his MSc in geography using OpenStreetMap. And so we got together and figured out this company and called it CloudMade. And then armed with this seed funding of—I think $100,000, we built a website and started to figure out how to get a tile server together. We hired a couple of key people in the community in London. And from there, I went and got VC funding.

That sounds like a good adventure. Was there a story there?—I know it ultimately failed but what did you do along the way?

The first thing to know is that 95% of these things fail and I had an expectation that it was going to be a failure. This is one of the great things about people who do VC. Normally entrepreneurs come into it very naïvely which is what your VC relies upon. You expect that you're somehow going to be the success, even the with overwhelming odds that you're going to fail. So yeah, we went and got VC. We hired lots more people. Ultimately, it failed because the incentives for success just weren't there. There we're a lot of incentives to move money around. But there wasn't really any incentive to succeed as an actual company with customers. And so, as soon as I realized that, I figured it was a better choice to go and go find something else to do, and I ended up at Microsoft.

Well, let's just hold off before we go to Microsoft for just a minute. You had this Linux Red Hat analogy. Was that analogy faulty in some or half-baked in some way?

No. I don't think so. Today, you see MapBox which is another VC-backed company doing exactly the same thing and burning even more VC money, going after it. I think the model was correct. I think the opportunity is still there. I think, just honestly, our timing was wrong—we were very early. There wasn't a lot of clarity around how exactly the chips would fall.

The other thing that was happening in parallel, remember, was the mobile revolution. The iPhone and then particularly the iPhone 3GS put maps and GPS in people's pockets. And so, that was the real opportunity for OpenStreetMap, not necessarily upending the old models, but being part of the new wave of the iPhone and then Android as well, of course.

And so, what you see Mapbox doing is basically providing map tiles as a central revenue stream to developers producing applications. That revenue stream didn't exist when we started CloudMade because the phone hadn't rolled out. It's partly waiting for the right technology-product-market fit. And partly that it took a long time to upend things, because if you look at the major sources of revenue for map companies, they tend to have very long lead times. It's in-car GPS or Yahoo, or Google, or other places where you had a requirement that the map data had to be of a certain quality, and OSM wasn't there yet.

The amount of money you could make from bus shelter posters is obviously dramatically smaller than if Yahoo or Google or General Motors or someone like that was going to be paying for it but you had this huge mismatch. It might be simpler to put it this way, that people think that deals are done a certain way when they're done another. A lot of these kinds of deals are done on golf courses with your sales guy. Unless you have those long relationships and a golf course then you're not going to get that same deal in the same way that today Mapbox can get—people's paying $5 a month or whatever for map tiles on the internet as a new market.

The point was to go through this process, learn about all this VC stuff, and end up living in America. That was my ulterior motive in all of this.

And so, how did the Microsoft job come about?

I'd been in Colorado for a while, futzing around with a bunch of different projects. Several different people at Microsoft reached out to me and said that it might be interesting to do something because Microsoft's map supplier was Navteq which is one of the two primary global data map suppliers. And Navteq had just been bought by Nokia, which back then was a global supplier of mobile phones. Nokia wasn't necessarily on the best of terms with Microsoft because there was some competition with Microsoft having its own mobile phone operating system and so on.

So Microsoft didn't have many options, because for buying navigation data there is an effective duopoly between two suppliers, Navteq and TeleAtlas, which had been by TomTom.

Alternatively, you could go drop several billion dollars on creating your own map, which is basically what Google did. Because the map suppliers know that you're in a tough spot. What they can do is raise the price on you, but it doesn't necessarily have to be the dollar amount. It could be that the licensing terms are onerous or the process is very long and complicated, or a number of other things. What happened, I think, is that Google just got bored with all of these and decided to go build their own map in order to have complete freedom over what they do.

Another thing is that as these mapping companies were trying to grapple with the internet, they tried to take their old business models and apply them. It used to be that a map page view, you'd have to pay some fraction of a penny for. So, if you had a user click to zoom on a map, that would be a new page view. And then if they zoom back out, that would be a new page view. So you're just racking up these page views and you have to pay money for each page view to the data supplier which is fundamentally restrictive on a user being able to interact with a map. Google didn't want that issue and neither did anybody else.

The third option for Microsoft was to use OpenStreetMap as the open competitor because at the end of the day, most of these guys are not really competing on the map data. Everyone's using the same map data. If Yahoo and Microsoft are both using Navteq map data, it's the same map data. Then they're competing by doing things on top whether it's tweaking your search experience having nicer colors, or whatever it is that you've convinced yourself is going to get you more customers.

Using OpenStreetMap data you would have more capital freed up to improve things. And so, eventually the world will probably shake itself out like that. And people will just use OpenStreetMap as a default platform. But at the time, OSM wasn't good enough.

Microsoft could spend some amount of money to improve OpenStreetMap as a bet, right? And so, they could do things like make the aerial imagery available, do some work to figure out GPS data--turning GPS data into speed limits, those kinds of things. To help improve OpenStreetMap to the point where it was usable. And so, there was a whole plan put together to effectively bring OpenStreetMap up to the quality required to replace commercial vendors.

However, Nokia went from a being a competitor to being a very strong business partner of Microsoft. Nokia took Windows Phone 7 and then 8 and 8.1 and, in exchange, Microsoft made a bunch of commitments to take Nokia maps, which had been Navteq, and use it more in key areas, in devices and in the operating system, and on the web and so on. That deal made complete sense from a strategic point of view, so there was no longer any need for OpenStreetMap. And in fact, OpenStreetMap was problematic for Microsoft because it was going against the grain of the new company direction. And so, because of that, I went and did other things.

It's one thing to bring OpenStreetMap to Microsoft but now, aren't you working for what essentially Open Street Map was trying to eliminate? Was there a perception that you went to the dark side here?

Not really. The perception was certainly there because Microsoft has a long history of being seen in the open source and Linux communities as very competitive. Today, that's just not really the case. Microsoft does all kinds of things with open source and is very supportive all across the board, actually much more supportive than many of their competitors in some ways. But there was this long history and so yeah, there was a perception in the community that Steve has become evil or whatever.

When you went to Microsoft?

Yeah. I mean, there's a long, long history of this. When I started CloudMade I was evil. They literally use the word evil. It's not shorthand for something. And when we got venture capital, I was evil. When I went to Microsoft, I was evil. So I mean, it's just the levels of evilness.

But I mean, you were the evillest when you went to Telenav, I would imagine, on this line of thinking, right? It just got worse and worse from that perspective?

No, most evil at Microsoft, I think. Yes, because Telenav takes the map data from those commercial suppliers and then does a bunch of work with it to put it into cars. So, they're in a similar position to everyone else in which the water's changing, And different customers want different maps. Customers just started asking about this OpenStreetMap thing, right? All these companies went through this process of becoming map agnostic. So yes, we take data from commercial suppliers but we also take data from OpenStreetMap. And we help improve OpenStreetMap, because if you're a company using commercial data, you don't have a right to improve it. You're not allowed to. That data is locked down. It's not like you can go in and just fix things like you can with OpenStreetMap. So, the project I worked on, we spent a lot of time and people and resources bringing OpenStreetMap up to the level that a consumer would want in the United States.

Did that work?

It did work, amazingly. We did all kinds of tests and metrics. We floated it. We allowed small numbers of people to use it, to see if they would notice that it's the same kind of quality. And we managed to switch over Scout, a consumer-facing turn-by-turn app to use OpenStreetMap in the U.S.

So, OpenStreetMap in the United States has enough data to do turn-by-turn navigation, and get you to an address where you want to go, properly?

Yes and no. There's three part to the map. There's the display part. So, you look at the map and it looks good. There's the geo-coding part. So, you need to be able to turn an address into a location. I type in '100 Main Street' and it has to know where that is. And then the third part is the navigation part. So, it looks good, you have the address but you need to know "Can I turn left here? Is there a stoplight there? What's the speed limit here?" All of those turn restrictions and other things that go into making a map actually navigable. Because, otherwise, a computer could route you the wrong way down a freeway, those kinds of things. Unless it knows that it's a freeway and that it runs in a certain direction.

OpenStreetMap has focused an awful lot of being a display map. And so, it's a great display map, but the geo-coding is lacking. And the navigable parts are missing too. The nice thing about Telenav is that through processing a lot of GPS data, you can extract a lot of those navigation items like speed limits and turn restrictions because you can look at a road and you can notice 100 people drive down it and most of them are driving at 55 miles/hour, then the speed limit is probably 55 miles/hour. You can look at an intersection and people make all kinds of turns at an intersection but they never turn right or something. Then you can say, "Hey, there's probably a turn restriction there so you can't turn right." By doing all of that you get the navigation data.

In the U.S., it's relatively easy to license geo-coding data. You can license the locations of houses, the house numbers and the street names and so on. And so, by doing those two things, you sort of use OpenStreetMap as the framework, and then you improve it a little bit with navigation information. And then when a user types in an address you use this third-party data set that you licensed. And by doing those couple of things, you're able to bring OpenStreetMap up to the same quality expectations. It's sort of like the last mile, because the community's done all the hard work of going out, mapping all the roads, getting their names and figuring out how connected they are and so on. And you're just doing the final 5% or 10% that turns it from a really good community project into something that's commercially viable.

It's funny because it's finally like you have a totally free map that works for what most consumers want to use it for, besides driving their cars. But the other not-free maps are actually free of cost to the consumer so there's an irony there.

Most of the profit centers in mapping just fell off the face of the Earth. Google led the way by making their maps very freely available both to the end consumer but also to developers to extend and put maps into web applications, and then later applications on your phone. And so, there's now definitely a consumer perception that maps are free. And really, all the costs are just hidden behind search display advertising, and other things. But it created an interesting environment where there's a lot of froth trying to figure out, "Well, what are we going to do if maps are now free?" It costs an awful lot of money to go collect map data and keep the maps up to date. Where does that money come from?

What is the answer?

It's a good question. I mean, it's multivariate, and it depends on the timescale we consider. But ultimately, there's been a long and fairly irrational chain of events that result in large companies, like Google and Apple and now Microsoft too, investing very heavily in mapping. They hide those costs behind whatever their core profit centers are. So, with Google it's advertising. With Apple, it's selling iPhones, with Microsoft, it's selling Windows licenses. And they take that money and have skirmishes around things like mapping, in the same way the United States and Russia have a bunch of money and just take turns in screwing over Afghanistan.

The question is how long are they going to be irrational about it? How can they turn maps from cost centers into profit centers? How can they turn them into something that people are willing to pay for? And so, there's been all kinds of attempts by all kinds of people over the years to try and figure out what people would pay for.

I think that the example a consumer would probably know best is offline mapping. The ability to use the map for free was great, but it only worked while your phone was connected to the network. Then you might have to pay some number of dollars if it worked when there was no network. Maybe you would travel to a foreign country where your phone didn't work, or maybe you're somewhere rural where the connectivity was pretty bad. And so, you pay for that as an additional service. But today, even that is now just free and part of the package.

It will probably be solved in two ways. The first is just crowdsourcing, because the cost and complexity of crowdsourcing is dropping dramatically. You know, when OpenStreetMap started, this was all very hokey and complicated and hard. But today, pretty much anyone can create crowdsourcing software on a phone. It's very helpful if you have millions of users. If you control Android or iOS and every time a user has their GPS switched on, you're getting a copy of that data so you can go build maps out of that very easily. So, the cost and complexity of actually mapping is going away fairly rapidly. That's the first thing.

And then the second thing is, I think that it's going the way of spell check. It used to be that maps were a thing, in the same way that in the '80s when web processors were new, spell check was a thing. It was a thing that you might have to pay for and you might have to license a dictionary. And you'd buy software specifically because it had spell check. But now, every text box you would ever use on any computing device anywhere just has spell check built into it. We have predictive text when you type a message on your phone for example.

People now, they don't think of spell check even as a feature. It's just the fabric of how computers work. Maps will be the same. We still have individual map applications but if you think about it, that doesn't really make a lot of sense. Why do I need a mapping application? What I really want is just want a map to be everywhere where the context demands it.

A good example would be you're trying to find a pizza place near you. Should you use the map application? Should you use a search site? Should you use something like Yelp? Or should you look in your contacts because maybe you've saved the pizza place in your contacts. When you're asking a simple question like "Where's the nearest pizza place?" And then having these different options that people respond to in different ways, ultimately, I don't think it's going to make a lot of sense.

We'll converge on some way of merging maps, search, ratings and your personal contact information and your interests. All of this will just become the same thing. Today, we just have this sort of hokey way splitting up the problem. If you're spatially-oriented, you go use the map application. If you don't know about crowdsourced ratings, you'll use Google. If you know about crowdsourced ratings, you'll use Yelp. And if it's your favorite pizza place, you have their location in your contacts. All of that stuff is going to go away and there probably won't even be a stand-alone map application unless you're someone who's just interested in exploring and playing around with maps in the same way that you'd use Google Earth. But in the future, it's just going to become part of the fabric like spell check.

Do you think there'll be separate map-siloed map databases? Or will that all kind of turn into you know, a common dictionary? A pool of data that everyone draws from? You used the word irrational. I think you were talking about accidents of history, but I think you're also saying that we've ended up in an irrational place. Is that what you meant?

Well, yeah. I mean, there's irrational amounts of money being spent on map data today that don't really make sense because the people making decisions around mapping are viewing map making as a business. Like, I have a car company. I need maps in my cars. I'm going to buy a map company. That's the logic that goes into it rather than going five levels deep and figuring out "Why do I need a map company? Where should the data come from? Does it need to be that high quality?"

A lot of these mapping efforts will go away as you get more feedback in the system. For example, the primary people who pay for maps today are people who buy in-car navigation. That supports lots of industries. It supports map data collection and so on. But it's going to become very problematic if self-driving cars happen because you don't even need to look at the map anymore. So, I don't need a big screen. I don't need to show you the map unless you're particularly interested in it. There might be a collapse in vehicles because if there are self-driving cars, you think "why do I need to own a car? Because you could just send your self-driving car to pick me up or I could use some sort of ride-sharing service.

And so, that's what I mean. There's a lot of change coming to the traditional revenue streams in mapping. A lot of the decisions are made without any particular feedback. So, you could look at on a sort of strategic level like, "Hey, I'm going to buy a map company. And then, I'm going to get my bonus and leave, and then someone else has to clean up the resulting mess." But also, on a very micro level, when you buy a car and it has a map in it, and the map is wrong. What are you going to do? I mean, right now, you can do nothing basically. You can switch providers or use your phone, but the die has been cast. You've already bought the car. It already has a the map in it. So, it's not like you can go and get your money back or complain to anyone. There's really no way for you to fix anything. There's no feedback.

That's going to change because we have this intermediary step between the cars we drive today and self-driving cars, which we call connected cars. Those connected cars will have always-on internet connections will need a higher perceived level of quality.

So if the car is always online, why does it have this shitty broken map? I'm now paying some monthly subscription to make all self-driving cars work but it's got a crappy map. That will put pressure on the people who build these data sets to improve the quality and drop the cost. The only way they can do that is to crowdsource the maps.

Has the traditional mapping industry collapsed or evolved? Has it just been washed away by this wealth of free data? Or have they figured out how to embrace the new and build new things?

It's a little bit of both. If you look where these guys have landed, there's really two sources of funding. There are a couple of large map data companies that are supported by vehicle navigation and other stuff. And then there are governments. The governments haven't really changed, and they don't need to change because they're governments. So, they keep doing whatever it is they're going to do and provide the same maps.

Eventually, there'll probably be some seismic changes there for a couple of reasons. One, if you're some tiny, little European government and you have a mapping agency, it's no longer particularly relevant that you have a map of France or England or Spain because your consumers are expecting maps that are pan-European or North American. And so, the government identity and role in funding—all of that is being questioned. But it takes 10, 20, or 30 years for them to figure that out. For the map companies, I think it's the same process, but it's just a lot quicker. And so, they try to innovate. They try and attach things like, "Hey, we're going to do self-driving cars" Because, theoretically, a self-driving car requires a lot of high-fidelity information. I mean, it's just deeply problematic.

There are things that they wave around and say, "Hey, we have a future business because of X." And then there's reality. They'd probably say, "Hey, we need much high-definition maps and that's what we're going to do." But when you step up the scale in mapping, you also step up the scale in the cost and the complexity. So, if you map all the freeways, that's one thing. And the freeways haven't changed much in 50 years. But you map all the secondary roads and tertiary roads and so on. And then you get a lot of change. And then you map all the residential roads. Residential roads change every year. And then you map all the footpaths. The footpaths change every week. They change every season.

And for a self-driving car, it needs to have incredible precision and detail in its mapping. The precision of detail is beyond the realm of what you can accomplish by just paying people to drive around and collect the information, partly because, driving around, they might be wrong. But secondly, it's just a ridiculous amount of information. It's going to change all the time. The solution thus far has been "Well, let's just drive every year." Or it used to be, "Let's drive every road, every year," but even that is unsustainable.

And so, what it comes down to is that the car itself is going to have to be mapping as it's going around. It's going to have to be very tolerant of change. So, if it expects a stoplight and there isn't one, it's going to have to figure out what to do. You know, what Tesla did is very clever, which is just focus on freeway driving because it's sort of the 90% of what you're doing. When you're going from A to B in the United States, it's typically get on the freeway, drive some amount of time and then get off the freeway.

The freeway is great because it's a very constrained problem; freeways tend to look alike from place to place. They tend to not change very much and have a fairly orderly set of rules about them like, "I'm going to stay in a lane. I'm going to enter at this exit and exit off that exit." Whereas the rest of the street network is just a big mess. And so, if you can't rely on just paying people to drive around and collect this information, and if the car has to be tolerant of this change, and has the ability to figure out these stuff on its own, then it might as well be mapping as it's going along and helping other cars around it with information and letting them know what's going on and fixing the infrastructure. But then, you also have to ask "Why do we have any of this infrastructure?" Because if you have self-driving cars, you don't need stop signs and stop lights anymore. And so, that entire positive revenue stream goes away, because if you're a car company like Tesla, why would you rely on a base map at all? Why wouldn't you just go build your own and have the cars automatically update it all the time?

And so, initially, I think what we'll see happen is that lots of people have their own maps. It used to be that there was just Teleatlas and Navteq. But, now, there's Teleatlas, Navteq and Google. And then Google bought Waze, which means that there's four maps. Now, there's OpenStreetMap.

I've got no information on this, but it's pretty conceivable that Uber would make a map, that Tesla would make a map. So I think, if anything, you're going to see lots more maps, you're going to see lots more competition, which means the price is going to drop. And then there'll be some consolidation because it probably doesn't make a lot of sense for GM to be making maps and Ford to be making maps, in the same way that "Why would they both make hub caps? Or why would they both make tires?" You would just rely on some third party that does this, but with a new model.

What's the next move? What do you think the right move for OSM is? What's the right move for you? What are your personal goals kind of around mapping in general? And how they overlap or not.

Steve Coast

I've always been interested in doing new things. OpenStreetMap was a new thing. Then CloudMade trying to monetize OpenStreetMap was a new thing. Theoretically taking the resources that Microsoft had and applying them to OpenStreetMap was a new thing. Each of these stages has led to boosts in the project. And at Telenav, what we managed to do by making OpenStreetMap commercially viable is huge. It means, that no longer is OpenStreetMap this laughable side project put together by a bunch of hippies. A lot of people in OpenStreetMap are still stuck in 2008, They think that it's just a bunch of friends getting together mapping. That's fine up to a point, and it's interesting. But for me, I've got to keep moving and go do something new because it just doesn't interest me to be solving the same problems repeatedly.

Also, OpenStreetMap is a take on crowdsourcing that's different from Waze. It's a take on crowdsourcing that's different from Wikipedia, both in its technology but also in its human approaches. When you talk about the future, it can be also about taking the components of Open Street Map and applying them elsewhere. So, what else can you go crowdsource? Given that I learned how to make conferences in OpenStreetMap, what other conferences could you go make? There are lots of other data sets out there that could be crowdsourced too.

Addressing is a big topic. So, OpenStreetMap requires addressing for navigation for consumers, so addressing is obvious.

Datasets of pictures of things are also obvious. A picture of a car and it says it's a car. A picture of a BMW and it says it's a BMW. You would be amazed at how those data sets don't really exist. And there are all kinds of uses for those data sets, for example in machine learning. There are entire industries built on different data sets that should just be open. For example, turning an IP address into a location—which is still very relevant despite being an old technology. That's still pretty much an unsolved problem.

So, if you use a laptop and it doesn't have a GPS in it, it has to figure out its location. One of the bad ways of doing that is using your IP address. And so, there are companies that exist that just to supply databases of IP addresses and where they are located—sort of reverse engineering the topology of the internet. And these companies charge a reasonable amount of money for that. And you find these data sets all over the place. There's StreetView and a company called Mapillary that are giving you a a sense of place via 360-bubble pictures.

It's interesting how many of those are really extensions of an underlying map.

Yeah. It's absolutely the case. But if you think about it, you should almost expect that because so many types of data have place inherent in them. A dictionary doesn't, right? It's just a big long list of words, in a sense. But you go beyond those simple examples like a dictionary and almost everything has place associated with it. You take a photo, it has a location straight away. The objects in them have geographical histories too. It's a picture of BMW, it was made in Germany. And so, it's actually, when you get into it, it's pretty hard to find data sets that aren't reliant at some level on location.

Is the map, in a conceptual sense, an operating system? I mean, is it just the lowest level that pretty much everything needs?

Yes. I mean, I'm struggling a little bit because, from a computer science point of view, the analogy falls apart a bit. But the basics are yes, you need some sort of underlying structure on which to lay things in order for you as a human being to make sense of it. But also, for the computer to be able to figure stuff out. Your phone has a GPS in it. You take a photo and it has a latitude and longitude, but the phone really needs to know a lot more. It needs to know, are you pointing north or south? What town is that latitude and longitude in? Those things are contained in the map, which we could call the OS for everything else.

Martijn van Exel

Bio

As a little kid, Martijn had dreams of changing the world. Then his grandparents gave him an atlas for his sixth birthday. After that he just dreamt of changing the map of the world. Skip to 2007 and a degree in urban planning later. Along came OpenStreetMap – a map of the world anyone could change. Dreams became a reality, and the rest, as they say, is history. Right now, Martijn sits on the Board of Directors for OpenStreetMap US, and changes the map of the world at Telenav, where he works tirelessly to bring OSM to new devices, people and places.

Favorite Map?

"I have a hard time picking favorites. I love the work of designer / cartographer Cameron Booth. Recently, he digitally restored the first map of the US highway system from 1926. This to me embodies the combined power of old world map making, modern technologies and open (in this case, out of copyright) data."

http://www.cambooth.net/project-1926-us-highways/1926_ushighways_2500px/

Interview

Who was Martijn before you found OpenStreetMap? Where were you? What were you doing?

I was just starting a career that had something to do with geospatial stuff. I'm not, by training or profession, a geospatial person. I was trained as an urban planner and I did some mathematics on the side. I became very interested in programming. I had always been interested in maps. So those things kind of started to come together. It happened a couple of years before I got into OpenStreetMap, when I got my first job as a GIS person at a company in the Netherlands, in Amsterdam where I lived at the time. I started following news around GIS and maps - digital mapping and OpenStreetMap started to come along, for me, around early 2007 I would say.

What was your very first interaction with OpenStreetMap?

The very first thing I saw was actually–I don't know through which platform or forum but I saw an announcement for a mapping party in Amsterdam. That's when I thought, "This is super interesting." I had held GPS devices a few times and played around with them a little bit but I was just interested in who these people were. I didn't know anything about the project yet but I thought like, "I'll just go do that thing" and it was a ten minute bike ride from home so that's when I decided "I'll have to find out what this is."

So you didn't think, "Wow, these guys are crazy."

No. Never. I mean, I never thought it was crazy. I had no expectation that it would go anywhere but I had a feeling that it would be at least an interesting group of people. For me, it was mostly about the people - like "who is interested in this stuff?" Like I said, I was just starting a career in this field. I was just curious about what kind of folks were out there who are interested in and actually making maps.

When you start a career in GIS, it's not so much about making maps but it's a lot about doing stuff with maps that other people make and you never really get exposed to the making part. I was always really interested in the making part but you don't really get taught that or exposed to that. So when this came along, I was like, "Okay, this is interesting. I've got to meet these people and figure out what they have to say and how they go about what they do."

So what happened?

So what happened? This was actually a mapping party. In the Netherlands at first they were throwing these pretty large mapping parties. They were over two days—a whole weekend. There was this guy, Zoran, whom I think you know. He had his own company in this beautiful historic building opposite the Amsterdam Central Station. He had the whole top floor and he invited just everyone over for a weekend of mapping. There were 15 or 20 people there which is much more than I thought. It was interesting because it was all about actually collecting. What I expected was people talking a lot and looking at strings and software and doing stuff with lines and polygons and points. But what happened actually is we got handed out GPS units that he got a hold of and we got sent away to the streets and mapping. Basically it was figure it out, and come back at the end of the day and we'll figure out how to put all this stuff into OpenStreetMap. So it was kind of a deep dive - a very interesting one.

So we went out in groups. At first, we had a quick look at the map of Amsterdam which was pretty much empty but for a few main features like the major roadways and a few water features but nothing much. Most of the neighborhoods were basically blank slates so I ventured out into my own neighborhood and collected some GPS traces. I don't think I had a digital camera but I took some notes of my GPS traces and that's it. And then we came back and then we worked with, I think, a very early version of JOSM that was already out by then, and we started helping each other out digitizing. I had a ton of fun. I didn't plan on coming back the next day but I did and I was sold, basically.

Were there any social activities involved?

Yeah. So we went out for dinner one night and drinks. Zoran had his company there and they had a foosball table and a projector with an Xbox and games. Everybody pretty much stuck around into the evening, which was kind of—I expected this to be like a few hours in the afternoon but it turned out like it went way into the evening. That kind of set the tone for a lot of later OpenStreetMap events that I both attended and organized. The social aspect was very big from the beginning. I don't know if I'm biased in that way but that was definitely an important part of it.

What were your next steps after that mapping party?

I started figuring it out at home by learning more about—at first, just the mapping and where these people were subscribing to, sort of the mailing lists - the Dutch one and the international one, to see who else was out there. I think I went on to the IRC channels pretty early on also just to get a sense of what kind of people were out there and had some early discussions about mainly, at first, just questions about do you map trees? How do you map stores? How do you map all kinds of stuff that I wanted to add? And then we continued to organize mapping parties. There were, I think, a couple more. I wasn't involved in organizing at first but I got involved the next year, I think, when I became interested enough to start being part of the Dutch community leadership and organizing parties and events also outside of Amsterdam.

Were there any changes in your work at this time?

No, not right away, not until a couple of years later. I was with this company that did mostly mapping and geoanalysis stuff for governments. I was in the research department so we were supposed to come up with new ideas of how we can sell this kind of stuff to governments and to private parties. So we really just started looking into open source which was becoming big in computing then, but was still pretty new in the GIS domain. People were not quite receptive to it. And then I slowly started to introduce OpenStreetMap as an alternative for, at first, just the traditional kinds of display maps and base layers. So when I started to talk to people "Hey, this is OpenStreetMap and we can use it."

If you work in a very traditional GIS business, it's not something that you can do very quickly. You have to have a big amount of patience, but they gave me at least some time to explore and to help, for example, organize the State of the Map Conference in Amsterdam which we had in 2009. I got a good amount of time from my employer to help organize that conference. They were receptive to the idea but there was not a lot of adoption at first, so that was pretty slow.

You said the first mapping party was in 2007. What happened over the next few years?

In the Netherlands, we had an early break where we got in touch with this company, AND, which wanted to donate the street network of the Netherlands to

OSM. And you were conducive in that whole effort. You talked with them and you made the initial contacts. And then we started actually making sure that this data made it into OpenStreetMap. Then very early on in 2007, the map of the Netherlands suddenly looked fairly complete. This was a pretty big process that I was actually not very involved in. At the time, I took some time off of work and being in front of the screen altogether. So the whole import of that AND data was not something that I was deeply involved with. But it was something that when it was done, I think, it made a huge difference in what the community grew into. So before that, there was this community of explorers – people who want to draw on a blank slate.

After that we started to attract a lot of people who were interested in using the data and also seeing how we can collect more data sources to make it even more complete. So instead of actually going out and surveying, it became more and more about how you can fill the gaps with more external data. Over time, that kind of balanced out into a lot of people still went out mapping while other people focused on ever more data imports.

So you see in the Netherlands, I think, a very interesting blend of pretty high-quality imported data. There have been quite a few imports since then. And also, a very active community of people who just go out and survey and map. I was afraid at first, when those imports started happening, that there would not be a living community, that it would not be a living map, if you will, because it would just be a dead repository of data that was also available elsewhere.

It turned out I was wrong. Actually, it turns out that an OSM with lots of imported data doesn't kill off the community and I was happy to be wrong. But it's something that I definitely didn't expect to happen. I mean, this whole community thing was pretty new to me. I hadn't really been involved in open source software or any kind of collaborative communities like that before. So yeah, the early predictions were mostly wrong. Where I thought like, "Okay, this will not—people will just go away and it will just be people from governments and third parties like commercial stakeholders who will be involved in OSM and no actual "on the ground" individuals mapping anymore." Yeah, that's all turned out to be wrong.

How do you think that debate has evolved to today, in the project about importing data?

I think, there's different perspectives depending on where you come from, right? To my mind, simplifying things a little bit, there's two camps, right? There's the camp of the people who have been part of the communities in countries where all these imports have taken place and I've been part of two. I've moved from the Netherlands where I just talked about the imports that we've seen there, starting with AND. And then later, a nationwide buildings database and land use. So we've seen a fair amount of imports in the Netherlands. And then I moved to the United States in 2011. Before I moved there, in 2008, the folks there had imported the census TIGER data which is basically also road network and of pretty questionable quality.

And I think, but I can't be sure, that a similar debate was going on in the OSM Community in the U.S. at the time of that import and perhaps also outside where people thought like, "Okay, if we do this, how are people ever going to be inspired to go out and survey if they look at the map and they see all these roads are already there?" And again, those people turned out to be wrong. Where in the U.S., also aided by I think good leadership and good community initiatives, we have a thriving community. I mean, it's still fairly small if you think about population of the U.S. being 310 million people and the number of active mappers in the U.S. month-over-month is more like 1000 to 1500. I mean, that's small but its' still people. There's a good amount of organizational initiative where we have like two to three dozen fairly active local groups and people organizing mapping parties. And we have mapathons four times a year. What I'm saying is there is a community that's very much alive in spite of, I would say, these imports in countries where there's never been any of that kind of activity in the importing data. Examples that come to mind immediately are Germany and the U.K. I guess where there have only been very small imports, if any.

People are much more careful now about accepting external data into OpenStreetMap. I don't know if that's because they think it will kill off the community. That's probably part of it and here's where I start being creative with just thinking, basically. And the other part of it is that like they have shown, to a large extent, that you can also evolve the map into a very high-quality product without external data finding its way in there.

What was the first international event that you went to?

Yeah, that puts a smile on my face, thinking about the State of the Map Conference in 2008 that was in Limerick, Ireland. I was just too late for the very first State of the Map 2007 in the U.K.—in Manchester. It was right after the time when I first became involved in OSM. I think it was the summer of 2007. My first involvement with the project was right about then in June, I think. I heard about it but that was a little early for me or I would've made it to that.

In retrospect, of course, I'm sad that I missed it, but the conference in Limerick, I went there with great expectations in the sense that I've met all these people online in the meantime from these different countries—from Spain, from Germany, from Ireland, from the U.K. Online, they're a pretty interesting bunch and I was looking forward very much to meeting them in person just to see how does this work on a global scale or even in Europe. It was back then, mostly European. I guess, on a European scale, European nations are very different culturally. What happens if you put all these people together in one project and put them together in one place and talk about maps for a weekend? It turned out to be absolutely fantastic.

Martijn van Exel

What were some of the highlights from that conference?

My strongest memory of the conference is arriving there. Of course, I knew no one there in person. I knew a lot of names and I knew whom to expect there. The conference was at this hotel at the edge of City of Limerick. It was a plain-looking affair. I came there, I was like, "okay, let's see if there's any others here."

Even before I checked into my room, off of the check-in desk, there was this kind of restaurant and bar area and there was a handful of OpenStreetMap folk. So I was sitting there and even before I had a chance to check into my room and put away my stuff, I was already sitting down for a beer with this dozen people that were all mappers that I'd never met before, and I was already drinking beer with them and talking about nerdy mappy stuff. I think that's my strongest memory, arriving there and becoming immediately immersed in that international community.

Okay, so you got to that conference and then, I guess, come back with positive memories. And then what did you do?

I became more involved in kind of the organizing side of things. So the Netherlands community started to grow also outside of Amsterdam. I think there were about 50 people who were like in touch regularly on the mailing list. So we started thinking we should organize more events and perhaps also maybe start raising some funds to help us connect with other communities elsewhere. I think we raised a modest amount of money like a few thousand Euros to support our events and we organized a good number of things around the Netherlands.

The Netherlands is fairly small. You can get anywhere in the train within three hours. We did events in most parts of the country and we got some sponsors. It all became a little more institutionalized to the point where we had our own server where we could do some stuff, mostly tile serving. I remember spending a long time setting up a tile server that was a little different from the—I can't remember what we wanted to do specifically that was different from the main OpenStreetMap files but they looked a little different because a few people felt very strongly about—I'm trying to remember what it was. So they wanted their own tiles for the Dutch server. And so, we did stuff like that also.

How formal was this? Was there a company or a non-profit?

It was informal in the sense that we didn't have any legal organizational structure. We did have a little bit of a formal structure to park that money that we got somewhere but other than that it was just basically a small group of people getting together from time to time or talking on the phone and trying to see what we can do to help, to get this community to thrive and grow. Of course, it became a little more serious when we started doing a bid for the State of the Map Conference in 2009 and we got accepted.

So Amsterdam was going to be the third conference and the first one—I think Limerick—was about perhaps 100 to 150 people. Manchester had been, I don't know, perhaps 50 to 75. I'm not sure. You would probably know.

It was 80.

It was 80. Yeah, so Limerick was probably around 125 to 150, perhaps. And with Amsterdam, we wanted to aim a little higher. And also, this was the first conference where we started to think about "How can we accommodate commercial interest in OpenStreetMap?" We started to get a little bit of attention from companies like the one where I worked. A company called Geodan in the Netherlands. But also other companies started to get interested in using OpenStreetMap data somehow or perhaps entering into a relationship where they could give and take perhaps to give in the form of money or data or other non-monetary ways such as sponsoring a server and perhaps then to take in the form of map tiles or map data that they could use for their own purposes. So we thought we needed to give those kind of commercial interests a place at the conference, both in terms of sponsoring of course, but also to give them a platform to talk about their needs. We decided to make the conference three days instead of two and have the first day be a kind of business day which turned out pretty nice as I remember. And we got a big break, the City of Amsterdam gave us an awesome venue to hold the conference for very little money so we could spend most of the money from sponsorships on niceties such as a herring stand at the conference. And we got to do it for three days. So this was the first conference where we had a full three days of scheduled sessions.

Can you talk about the herring stand because that's something that I remember?

We wanted to give this conference a little bit of local flavor, right? So what is the Netherlands known for? Perhaps not a lot but we like to think that people enjoy our fish and seafood because we're a seafaring nation. And so we thought we'd introduce this herring stand. The Dutch love to eat raw, cured herring. The way they eat it by gutting the herring and picking it up by the tail and lifting it above their mouths and eating it whole. We thought we'd introduce the OpenStreetMap Community to the Dutch way of eating herring.

How was that received?

Very mixed. What's surprising to me is most people actually liked it because it is a little bit of an acquired taste. It's like a lot of Dutch treats. I don't know if you've ever heard about drop—the black licorice the Dutch like a lot. I know just as many people dislike that stuff with a vengeance, as I know people who love them. Dutch cuisine seems to be kind of divisive in that way but the herring was received well. And I think also what helped a lot was that the sales people who came in traditional Dutch clothing to help complete the picture of the Dutch fisherman selling herring at the market and doling it out to our poor mapping community.

Was there any Stroopwafel at any point? Because that's not going to be divisive, is it?

No. I just handed out Stroopwafels at a party here at a friend's house in Salt Lake City last weekend and they were gone before I could even turn around to look at them again. So no, that's not divisive at all.

Are there any other highlights from the Amsterdam conference?

This was also the first conference that was connected, I would say. I think, in Limerick, we had some kind of internet connectivity but there was not really a thing that you did back then. In 2009, that all started to change and people started to expect internet connectivity. So that was a bit of an interesting challenge because as nice as that venue was in Amsterdam, they wouldn't guarantee more than I think 25 or 50 consecutive connections at the same time with their Wi-Fi network. So we said, "Well, we're sure not everybody's going to bring a Wi-Fi capable computer" and smart phones were not even really a thing back then, I guess. It turns out that everybody did bring their computer and they brought more than one, so the Wi-Fi network was clogged ten minutes after the conference started and it grinded to a screeching halt. So we were scrambling to find a solution for that. We actually ended up calling in this company that had mobile Wi-Fi access points based on a WiMAX that basically used the cellular network to generate more Wi-Fi capacity. We asked those people to come in on a Saturday and put some of those WiMAX access points in the conference to alleviate some of that trouble. It magically also worked and I think no one even actually noticed pretty much. So that was good. It cost a lot of money though. I remember that putting a good dent in our budget but everybody was happy so that was important. So we underestimated a little bit how important connectivity had become.

I think it was also the first conference where we had a Twitter Wall or a TweetWall where you have this big TV. I think it was a flatscreen TV where we could see all the tweets that had certain hashtags, #sotm or something like that. So you could see what everybody was tweeting. And this was 2009, so Twitter was pretty small back then but I guess it was already pretty well established in the OSM community.

And it was also the first conference where we did video recording and even tried to do video streaming. I think that miserably failed. But we did manage to record some of the talks and put them online for later perusal. That's something that we have done ever since, I think, for State of the Map conferences. And you see, in general, that the reach of those recorded videos is about ten times the audience in the room so it's definitely worth it. I think we made some good steps to kind of set the tone for future conferences. And in Amsterdam, I had a lot of fun doing it. It was a learning experience for sure but it was fun.

Around the same time, were you doing anything technical with OpenStreetMap or was it just sort of organizing events like State of the Map and mapping parties and mapping?

No. Not from my side so much. I mean, I've always been interested in the technology and I definitely was trying to understand it more. But for me, actually doing something that involves making technology or contributing to the technology that made up OpenStreetMap or extending that technology, that was not really something that I was actively involved in then. That came a little later.

Okay. So then after the conference, what were the next steps in your OpenStreetMap life?

Yeah. I think, after the conference, I took a few weeks off of OpenStreetMap because I had no idea how exhausting organizing a conference is. When you go to a well-organized conference and you've never actually organized one, it's hard to appreciate how much work goes into having an event that has a few hundred people and a program that needs to run smoothly, how much that takes. I was basically exhausted. So in the few weeks, right after the conference, I took some time off. I was kind of off the radar. And after that, I think I started to get more involved with the technology. I remember visiting the Hack Weekend in London. I think that was after State of the Map in Amsterdam. I met a lot of the people, not for the first time, but I met a lot of the people who are really behind the technology of OpenStreetMap.

Can you talk about what is a Hack Weekend and how's it different to a mapping party and the conference?

Yeah, so that's what I started to find out. I had no idea what a Hack Weekend was so I was curious. I didn't quite know what I had to contribute but I was interested in learning what it was. So it turned out to be very different. The mapping parties that I went to, that's where I expected people to look at screens and instead people went out and mapped and took a lot of notes and were involved with GPS traces.

Hack Weekend, on the other hand, turned out to be something different altogether. I came there and everything was like—I think you picked me up from the train station and we came into a room where people glanced up, but they were involved with either very technical discussions or looking at code and/or data on their laptop screens. It became a little more social afterwards but I remember entering that room and everybody like acknowledging me. I introduced myself but after that everyone was pretty much back to work. So that's actually where people get together to not only do a lot of the work that makes OpenStreetMap tick but also have those kind of deeper face-to-face discussions about the future of how the technology that drives OpenStreetMap should emerge like newer versions of the

API, database schemas and the future design of the website. So those are things that you can of course you can hash out with all that stuff over IRC or mailing lists or GitHub these days, I guess, also.

But these Hack Weekends turn out to be insanely valuable, not even so much because you could sit down and do a lot of work but more because you have all these minds together that think about OpenStreetMap for much of their waking life and about the technology that makes it work. You get them in one room and talk and share their ideas. That is really what moves the project forward on a technological level. So these things, I learned then are very important to how OpenStreetMap functions on the technological level. And they're fun.

I didn't understand the first thing about what a lot of people were saying. It was like discussions about Rails which I know nothing about. And others were talking about Postgres databases that at the time I knew very little about. And there were people coding in C++ for conversion tools that I really don't understand anything about. So it's just like most of what I did was just listen and try and pick up a little bit. I understood the value of it so I started organizing these things myself, but I'm still awestruck by the amount of knowledge that goes into OpenStreetMap. I can only scratch the surface of most of the topics they were talking about.

Who was paying for all of this? Just out of pocket? You're just flying around Europe doing this or what?

Specifically for me, I think, to visit that Hack Weekend. I think I paid out of pocket for that but in general—like when I started organizing these things, we started raising some funding. So, for example, I organized a Hack Weekend in Germany, just over the border from the Netherlands. So in Western Germany in a city called Essen there is a place called the Linuxhotel, and that's an interesting place. It's kind of a big mansion in the woods, very, very stately big place. And it's basically a convention center for open source seminars where people pay a lot of money to learn about open source technology. But on the weekends, they open it up to groups like OSM to have their events for very little money. And so we were able to get in there and I was able to get a few hundred Euros to cover food and drinks and pay most of the people's room and board, so people could just basically attend and just pay for their own travel and then nothing else. It doesn't really take a lot of money if you're creative. People are prepared to spend a little bit of their own money to travel around and attend these things because these folks are generally super motivated to get together and talk about OSM technology. It was funny. I organized this thing together with my German-Austrian friend Jochen Topf.

He knew about the Linuxhotel so we put it together. I was there. I felt like I was co-hosting this so I should know what's going on. So I tried to learn from everybody what they were working on but it's like there's so much going on in OSM. These were not only people working on the core of OSM but also in infrastructure and also building tools—or building new mapping style sheets for example to highlight one person's work on an outdoor style sheet. There's so much stuff going on that its impossible to keep track of.

Was that the one that Nick and I went to?

No. There were a few Germans. It was like ten Germans and three or four Dutch folks who were there.

I'm trying to figure out how do we get from there to Martijn's in the United States? And how do we get from there to OSMF-US and what's the progression?

It was pretty sudden that my personal circumstances changed and I was going to move to the United States. I was still involved with OSM Netherlands. I was organizing monthly geobeers or OSM beers in Amsterdam which were pretty well-attended and were always a lot of fun. At that time, I wasn't doing a ton of mapping myself but just kind of enjoying organizing the local events that I did. And people were talking about a new wave of imports and I was kind of unhappy about that. So I was still enjoying the social aspect of OSM very much but I was, at the time, not particularly super-involved on the mapping level except for my own little neck of the woods in Amsterdam where I lived.

And then in the summer of 2011 I moved to the U.S. to Salt Lake City which is basically in the desert in the western United States. The first thing I did was to check out if there was an OSM community there and there wasn't really. There were people mapping already. There were definitely people mapping. I could see that the map was decent-looking but no one was really trying to organize it. I thought "yeah, well, that's fine but I wanted to meet those people", so I started a meetup group in Salt Lake City and started to do mapping parties and also events where I talked about OpenStreetMap. I went to the University and did a class on OpenStreetMap in the Geography Department and it started spreading the word a little bit and talking about OpenStreetMap as much as I can. I remember going to the State GIS Department which is here in Salt Lake City and talking about OpenStreetMap and seeing if they're interested in doing things together like data sharing and perhaps using OpenStreetMap for some of their projects. So yeah, I started kind of building a community of interested individuals and also professionals over the course of that year.

I just started doing basically what I had done in Amsterdam but then my move coincided also with that State of the Map 2011 conference which was in Denver, Colorado which is really right next door. It's like an hour flight from Salt Lake. I came to the US with basically no job or no prospect of a job because my wife was the one who got a job here in Salt Lake City and I was basically just tagging along and figuring out what I would l do next. So I thought, "Well, let's at least definitely go to the State of the Map here in Denver and meet more people. Who knows what will happen?"

So in August 2011, I went to the Denver Conference which was pretty big. I can't remember how many people there were. I think about 300 people so it was a definitely good OSM vibe in the US. There were a lot of people from all over the world but it was I think probably more than half from the US. I had a good time at that conference and it kind of rekindled my excitement in becoming more involved with OSM again beyond just local community organizing.

So you go to the Denver Conference and then from there somehow got involved with OSMF—US Do you want to talk about exactly what OSMF—US is and then how you got involved?

I started talking at the conference with people who had been involved in starting the United States chapter which is basically the local OpenStreetMap organization here in the United States which was much more formalized already than it ever had been in the Netherlands. In the Netherlands, we always had been this loosely tied bunch of folks who were interested in OSM and got together at regular intervals. And here in the US on the other hand, there was a board and it was an incorporated organization and there was like a process, elections and that sort of thing. There was already stuff in place that I never seen but I thought, "Hey, well, if there is an elected board, why don't I try and be part of that." It's like I've done this kind of stuff before so I expected it'd be fun to do it.

I had no idea what I was getting into because I had no idea how big or small the US community really was. But sure enough, I got elected. I think, in part, because after moving to the US I think the first thing I did was of course get much more active in the talk mailing list – the Talk-US mailing list which is specific mailing list serve for the United States community which was fairly active. I started just asking a lot of questions and announcing that I had started a group in Salt Lake City and I guess people started knowing my name and name recognition leads to being elected as a board official in the United States. So I've been serving on the OpenStreetMap US chapter board ever since really, as the secretary first and then two-terms as a president. And then again, now again as a secretary.

Do you want talk through a bit more depth, getting elected and what you worked on and what you achieved?

One of the things that we realized fairly quickly is that there was definitely a US community, but there was not a lot of local groups or people actually getting together like I knew so well from Europe. So one of the first things that I think we started doing was trying to see if we could encourage folks to get organized and get local events going. There are a few ways we did this. One thing is just lead by example, right? So I did it in Salt Lake City and I shared my experiences like just on a personal level, you don't need to be on the board for that.

But then we started also organizing every three months. Once a quarter, we did a mapathon which is basically just encouraging folks to organize a local mapping party on a certain weekend. John Novak was instrumental in getting that going. We picked four weekends a year and then we said like, "Okay, put your city in a Wiki and just announce it locally and just get together with a lot of folks and map." And sometimes we'd pick a theme. Most of the time it was just getting together and mapping and getting to know each other, doing things fairly simple like that. I mean it's basically not a lot of work involved. It's just basically announcing that there is a thing called mapathons and that you can all participate. It turns out that that encourages a lot of folks to actually go ahead and organize something local they otherwise would not have done. So that worked fully well.

And then we started also doing that series of State of the Map—US. conferences, starting in 2012. Well, this was actually started before I moved to the U.S. There was the first one in Atlanta in 2010. Unfortunately, I never made it to that because I was still in Europe then. That one was the first State of the Map U.S. conference. Fairly small but, from what I hear, it was a lot of fun. I think you were there. I wasn't. And then in 2011, we had the international conference in Denver so we didn't do one that year but we followed up in 2012 in Portland, Oregon. And then in 2013, in San Francisco. And then in 2014, in Washington, D.C. This summer in New York City.

The U.S. conferences have been, I would say, fairly successful in terms of numbers of people and also the numbers of people from outside of the United States that we get there, and they've been a lot of fun to do. I think they really helped to solidify the idea that we have a community in the United States that is all working together to make a great map. The other end of the spectrum is these conferences. On the one hand, you can do these very small things.

Another example of things that we did is what we call mappy hours. I do these every other week on hangouts where I just send out an invite to IRC and the Talk-US—the mailing list serve and to other social networks that I kind of grew up with. Twitter and whatever. And then I say, "Okay, let's get together and just talk about OpenStreetMap and whatever you want." And so, we get together every other week for an hour. Usually, they're fairly well attended. We usually have a couple of people that are new to OSM and they just want to know more.

From my personal experience, it's just fairly powerful that you can actually talk to people live instead of interacting through an asynchronous mailing list or like other internet forums. If you actually can go online and see folks and ask questions and get answers right away. I think, for some people that works much better. It's another dimension of community building that I think is important especially if you have a country with the geography of the United States where everybody's so spread out. We're still one country, one nation but it's hard sometimes to feel that if it's so hard to actually get together. Again, from my perspective as a community organizer in the Netherlands where it's so easy to get people from all over the country to get together in one place even for just an evening. Here, there's no such thing. That was my main motivator for doing these mappy hours where you can just still get a feeling of being close and being connected in a way that internet forums and mailing lists and all of these other platforms cannot provide. I think it's a combination of all these different things that makes the community in the U.S. feel quite connected even though we don't see each other all that much.

So somewhere along this timeline you got technical, right? When and where did that begin and how did it end up with things like Map Roulette?

I'd been interested in the technology that drives OSM and the technology that other people were building to make use of OSM in different ways for quite some time like I said but I'd never really been involved in building stuff. And then came this period where I had a lot of time on my hands right after I moved to the United States and I didn't really have a full time job yet. So I thought, "Well, this is the time for me to get more involved with the technology that drives OpenStreetMap, and what can I do to kind of enhance that?" So right at that time, the license change was happening.

I don't want to go into the details of that too much but the end result of that was that the OSM community, after a very lengthy process of debating and discussing decided to change the license from Creative Commons CC-by-SA to a new license called the Open Database License and we had to get everyone to agree to this change. And to change the contributor terms that everybody agrees to when they sign up to OSM. And we were not able to reach everyone in that process, meaning the entire OpenStreetMap community. Those whom we were not able to reach, their edits needed to be removed from the system. This was a huge task but it was again like some of the technology wizards that came up with intelligent scripts to pick out all the data that was tainted by this license change because of all these people who had not accepted or we couldn't reach. And then we ended up with this map. I can't remember exactly when this was but I think it was late 2011 or early 2012 where we all of a sudden had these gaps where people who had not agreed to this change had particular data that was now gone. So then, for me, that was a driver. That was something that I felt like I could help do something about.

So this is when I started thinking about "What can we do to get this data back, right?" So I figured out a way to indicate to people where all these roads, which was what I was mainly interested in, that had been there were now deleted and asked them of course not to copy them back in because that will legally be very, very murky waters but to basically show them "here's a road that was deleted by this license change process and we want it back, so please go in to editor and fix it and see if you can look at the aerial image and see where that road is supposed to be and re-draw it and make sure that the map is complete again."

And then I thought, "Oh well, if we want to do this like 90,000 or 100,000 times in the U.S. alone, that's not going to be fun for anyone to do." So I was thinking about like what is a good way? How can we try and make this fun in a way? How could we make this into a process that instead of making people unhappy, to make people happy and excited about OpenStreetMap and about contributing?

So what I came up with is a predecessor to what later became MapRoulette. It's basically, you take all these things that are missing from the map and you know the places where that data was deleted and you put them in one big database. And then you pick them out one by one at random, right?

So once someone logs into the website—I can't remember what it was. I think, it was called the Remapatron. You go to the Remapatron. You get pointed at one specific location where data is missing and say "Please fix the map here." And then you go fix the map and you return to the Remapatron and you get another place where the map is missing. That kind of random but predictable sequence turned out to be very addictive when people started fixing the map at a rate that I've never expected. I'm starting to remember the exact number but I think we had about 70,000 or 80,000 missing roads in the United States and we've managed to fix that over the course of a few months. I can't remember if it's three, four, or five months but it was something around that time and this was something that we otherwise would never have achieved.

With my fairly limited development skills I thought, "Well, I can actually make a difference to this project in this community by doing something that's fairly simple to do." It just looks at contributing to the map or fixing the map in a different way that no one else has thought about and it turned out to be a success. I tried other things that were not as successful but this turned out to be a good way to help people find stuff to map. And in a map that's getting ever more visually complete, that can be frustrating for folks to not see anything of use to map. And if you can get them some random stuff to fix, that actually turns out to be very, very addictive and very encouraging, and people started to talk about this project. Someone else, I think it was Ian Dees—one of the other early U.S. contributors, came up with the name Map Roulette and registered the domain. And we started thinking about—

Was that based on Chat Roulette or something else?

You'd have to ask him. I actually don't know. I don't know if Chat Roulette existed then but it's a name that immediately struck me as it covered what it did so I thought it made a lot of sense so we had that domain. And then we started thinking about "What else can we do along the same lines to help, to get people to fix the map?" So we came up with other types of areas that we knew existed but were very hard to see and very boring to fix. And so we came up with this connectivity area where roads are almost touching but not quite and they're very hard to see. It turned out to be like 80,000 of them in the U.S. And with Map Roulette, we managed to fix them again in three or four months.

Since then, it's kind of become a staple of the OpenStreetMap editors or mapper toolkit, if you will. And since that time, I've opened up the Map Roulette to basically anyone with a little bit of perseverance that can make their own Map Roulette challenges and have people fix them. So we have map challenges from Italy, from Luxembourg, from Romania. We're looking at one from Canada. So there are all these people from all over the world now using Map Roulette as their platform to get stuff fixed in OSM. I think that it's turned out to be really powerful. I'm glad we could add that tool to people's mapping toolkits, if you will.

So why isn't Map Roulette just built into OpenStreetMap?

The simple answer is because I never really made an effort for it to be part of OpenStreetMap. I think it's been living in its separate existence fairly well. But it could be much more visible if it were built right into OpenStreetMap. It's just something that requires quite a bit of perseverance and a little bit of elbow grease to make happen and that's just something that I haven't done. It could not easily be done. It could be done but it just hasn't been.

Okay, so what are you working on today?

So stepping back a bit to 2011 conference in Denver, obviously I was looking for some form of employment. I didn't really expect it to be directly OpenStreetMap-related but one of my thoughts was "We're going to the conference not just to meet fellow OpenStreetMappers from the US that I have never met before and make new OSM friends, but also to see that OpenStreetMap had become much better known in kind of the broader professional geospatial community." And I expected there to be a fair number or people who might be able to connect me with–well, I don't know, perhaps people who are looking for someone with some geo skills. And it turned out that I met someone from, what turned out to be my employer and is still my employer, Telenav at that conference.

Actually, the story's kind of funny. One of the organizers at the conference asked me to be part of a panel. They wanted a few different mappers from around the world. I think I was one of the few people from the Netherlands. So they ended up asking me to be part of this panel, a panel discussion about: What is the OSM community like throughout Europe? What are the different perspectives on the project from different parts of the world? So everyone was asked to introduce themselves and I said, "I'm Martijn. I'm from the Netherlands. I just moved to the United States a few weeks ago and I'm looking for a job." And this person, Ryan from Telenav, he heard that and after the panel was over, he approached me and said, "Well, can we talk?" So I ended up doing some contracting work for them at first. Kind of trying to help them figure out what OpenStreetMap is. They had just started looking into it. And after about six months of doing that, they hired me full time and I'm still with Telenav looking at OpenStreetMap from a very different perspective than I ever have really which is "How can you make OpenStreetMap into a product that is ready for all kinds of navigation solutions?" to put it broadly.

So over your years of involvement in OpenStreetMap, what's the absolute highlight?

I think that first mapping party. It was so powerful. Most change in your life happens pretty gradually and you don't only really appreciate it when it's happened but getting in touch with OpenStreetMap for the first time was such a powerful event. I'd never really expected it to affect my life in the way that it ended up doing, but I did immediately know that this was something that I was going to be spending a lot of time with for the foreseeable future, at least. Like meeting those folks, it was a combination of the technology and the people that was so magical to me because then I've never really been involved with open source communities.

Richard Fairhurst

Bio

Richard is a cartographer, writer and developer living in the rural Cotswolds, England. He specialises in slow forms of travel - canal boats, formerly as editor of the UK canal authority's website and of Waterways World magazine; and bicycles, through the website cycle.travel. He is a church organist and occasional choir director with an unrealistic fondness for the far-too-difficult choral works of Herbert Howells.

Favorite Map?

> *"The Vale of Evesham, UK (http://www.openstreetmap.org/#map=16/52.0809/-1.8872), just 25 miles from my home. I'd got used to knowing all the mappers in the area - but then I idly browsed over here and noticed this incredibly detailed map, with every footpath, hedge and stream mapped to a level beyond even the Ordnance Survey. This, for me, was the moment that I realised OSM was unstoppable."*

Interview

Who is Richard Fairhurst and who was Richard Fairhurst before he found OpenStreetMap?

Who I was: absolutely not a developer or programmer. My day job was, and to some degree still is, writing for magazines.

When I was a spotty 15-year-old, I got started by writing the technical advice column for a magazine about Amstrad home computers, if you remember those—they came after the Sinclair Spectrum. Then I went on to a magazine about boats, canals, and rivers, which had always been an interest of mine, and then to British Waterways who were the people who actually ran the canals here in Britain.

Because I could still knock a computer program around in a fairly rudimentary fashion, I ended up coding the mapping for the British Waterways website. So, when "amateur cartography" started taking off on the internet, it was something I was interested in—then when all of a sudden, this OpenStreetMap project springs up, I think, "This could be fun to be involved in."

How did you first find out about the project?

Richard Fairhurst

With a bunch of friends, I'd had a similar idea for a collaborative mapping website which we called Geowiki. Back then Wikipedia was taking off, and the idea of wikis was quite big. We had this idea sitting around the bar on a ferry going to the Hebridean islands: I got home and had a quick play with building it.

It got a bit of attention, not too much. It was very primitive in coding terms, but the idea was there. Someone saw it and said, "Hey, you know, this is quite similar to what people have just started doing with OpenStreetMap. You should get in contact with them." So I subscribed to the OSM mailing list. That was probably the first mistake!

I can't resist going back to the narrative of the ferry to the Hebridean islands.

We were a bunch of friends who knew each other from university, not that many years beforehand. It was a pretty rough sailing back from a week's holiday on the Isle of Colonsay. And so what do you do on a pretty rough sailing, obviously, is you make your stomach even more vulnerable by going up to the bar where you've got all the big bearded Scotsmen working their way through all the whiskeys one by one.

We were chatting and thought, "You know, we've seen all this interesting stuff on Colonsay. It'd be quite fun to put this on a web map. Maybe we should build a site to do that."

Consumer GPS units were starting to be a thing then—the yellow eTrex was the in thing back then, about £150. So we thought, "If we want to build this map, all we've got to do is get an eTrex, walk around with it, and we've mapped some roads." Because we had the same experience that you did. There was no alternative to building your own map. You couldn't phone up the Ordnance Survey and say, "Hey, can you give me some of that data?" because they'd say, "Yes, definitely; that'll be £10,000 plus a support contract, and please sign here to say that you indemnify us from absolutely anything that might ever happen." That's not going to appeal to a bunch of kids sitting around a ferry back from the Hebridees coming up with a fun idea.

How did you get a job writing about computers at 15?

Before we had "open source", free-of-charge software was called "public domain"—a familiar phrase from many an OSM licensing argument! It was just software that hobbyists had written and wanted to give to other people. They didn't market it: it was just, "Here's something I've written, take it."

I wrote a bunch of it for the Amstrad computer. Some of it was quite decent: I wrote a simple desktop publishing program that a lot of people used. And at the same time, I wrote about it in some of the fanzines – enthusiasts' photocopied magazines.

At the same time, the news-stand magazine about the Amstrad – called *Amstrad Action*, which is a great name – needed a new technical writer. They'd seen my stuff in some of these fanzines and thought, "Well, you know, this guy seems to vaguely know what he's talking about and he can string a sentence together. Let's hire him." Which was a useful bit of extra pocket money in the late 1980s!

Which Amstrad?

The Amstrad CPC 464—this thing with these incredibly brightly-colored keys. 3.3 MHz processor, 64k. So really, really quite basic stuff.

What was the thing that came later that had Gem on it?

Oh, that was the Atari - the Atari ST.

But I thought Amstrad shipped a computer with Gem later on?

Yes, they did an early PC that was before Windows became universal. I presume Windows was on the way up, so Alan Sugar does his research and says, "Everyone's using Windows, that'll give me a cheap deal on Gem."

Back on the CPC, I had my first venture into all this mapping stuff. Do you remember AutoRoute?

I do. Bought by Microsoft.

Yeah, developed by Nextbase and then Microsoft bought it. I'd seen that on people's PCs and I thought that was quite fun. So I decided to have a go at writing a version of that for the 128k Amstrad. You didn't have much memory for the full routing graph, but it worked quite nicely: it had all the A roads of Britain. You could say, "I want to go from London to Edinburgh" and it would churn away for a minute and say "Go up the A1". That was my first venture into computerized mapping.

And again, the problem then was where to get the data from. So I got a road atlas and laboriously measured all the distances with a ruler. "Seven miles from Oakham to Uppingham. Let's have a little link in the graph going that way." I probably still owe the Ordnance Survey about 75p in royalties for my 1989 Amstrad routing program, I suspect.

Give me a sense of how Geowiki worked.

There were two elements to it. One was the idea of uploading GPSs to make a base map. You'd send us a GPS trace and say, "This is so-and-so road." And we'd add it to this great big Illustrator document which was a map of Britain, which then got boiled down to JPEGs to make a browseable map.

Over and above, users could put in little icons to say, "This is a café here, or this is a restaurant". Very freeform: "This is what category it is, this is what the name is, and here's my comments on it."

The whole technical side of it was incredibly basic. Back then, I had no idea what SQL was, so there was no database behind it, just a load of flat files. It would've fallen over if more than three people looked at it at once, I suspect. But it was a good playground. The site was quite nice by the standards of 2003 web design: it didn't necessarily look like a hobbyist project.

Were you a cartographer at the time?

I'd been working for magazines, and when the magazine needed a map, I would always be the one who drew it. I then built the mapping for the British Waterways website. So although I didn't have any training as a cartographer as such, I'd been doing it on and off for a few years.

I knew Mary Spence, who was Chairman of the British Cartographic Society, because she ran a company called GeoProjects who made canal maps. I learned quite a lot from her: I'd be making a canal map and she'd say, "Well, okay, but it will be better if you did it like this or if you put this on here, if you moved the labels around."

So it was learning on the job?

Absolutely.

Would it be fair to call GeoWiki a sort of MVP?

I'm not sure I'd ever claim it was particularly viable! But it was a proof of concept of the idea that normal people can build a map just by contributing their knowledge – which is essentially the same as OpenStreetMap. The barrier to entry to the website should be just that you know something about this place, and we're going to make it easy for you to come in and enter your knowledge in a way that other people can access.

So what were the major differences between GeoWiki and how OpenStreetMap operated at the time?

GeoWiki was probably the best part of a year and a half before OpenStreetMap, so there wasn't that much overlap in time. But essentially the main difference was that OpenStreetMap was a more collaborative effort to build the base map. Everyone could edit OSM directly, whereas in GeoWiki, you sent us the GPS trace and we put it on this big map. That worked reasonably well for a map of the main roads in Britain, but clearly it wasn't going to scale up to a complete road map of the world.

Within two or three months, GeoWiki had a slightly coherent map of Britain's main roads, which it took OpenStreetMap quite a long way to get. But on the other hand, OpenStreetMap had a design that would scale up to cover the world.

It's as if one was the cartographer's approach to the problem, and the other a programmer's approach. And they're both the simplest ways of going forward from those points of view. Do you think that's fair?

I think that's spot on.

Were you aware of FreeMap, Nick Whitelegg's project?

No. The first time I heard about that was on the OpenStreetMap mailing list, I think. I remember London Free Map, which was Jo Walsh and Schuyler Erle's project: they had a mapping event around Limehouse in East London. I think that might have been the first time I'd got the impression that there were other people who were having similar ideas.

It's interesting because they started London Free Map after seeing OSM and deciding that it was too simple and what you needed was GIS protocols going back and forth.

I've chatted with a few people about this over the years. We came to the conclusion that it's not really the technical stuff that made the difference with any of them. Sure, the technical solutions all had their advantages and their disadvantages. But the thing that made OpenStreetMap succeed was this community coalescing around it early on. You went out and did the grunt work of promoting it. Sure, I could build something in my bedroom and Nick Whitelegg could build something in his bedroom and that's fine, but it doesn't go any further. But you went to Dorkbot and events like that and would say, "This is OpenStreetMap. This is a project that you should be involved in" – the whole evangelist thing.

That was the biggest difference between OpenStreetMap and the various other projects: you went out and sold OSM. And for the rest of us who didn't particularly have any great ambition to go out and sell things, it became obvious that this was a great project to hitch our flags to, because it looked like it might be going somewhere.

I don't like hearing that because it suggests that to go do something big again would require all that annoying running around the conferences!

Once you're in a position where people are listening to you, that's easier. But back then, who were we? We were absolutely nobody, trying to upend the entire cartographic industry. Ok, that sounds too much like the whole revolution narrative, but we came from nowhere and it did require that sort of effort to begin with.

Early on, when I was doing my talks. I purposefully picked large enemies because it's very convenient and it gets people to rally around. Luckily, they were kind of large and evil.

The Ordnance Survey was a pantomime enemy back then. And it helped that the Guardian, for example, came along and say, "Hey yeah, we think the Ordnance Survey are bad guys as well." These days OS are the good guys in a lot of ways. But back then it made the narrative easy to understand for a lot of people.

Why don't you talk about your early involvement in OpenStreetMap, who you met and how you got involved?

Jo Walsh had e-mailed me because she had seen GeoWiki and said, "Hey, you should have a look at OpenStreetMap." So I did: I signed up to the mailing list and said, "Hello, this is me. This is the sort of stuff I'd been working on. Really good to see that there's a load of other people here."

I hung around the fringes for a while looking at the mailing list, occasionally weighing in and suggesting things in the way that you do with mailing lists. And what really started to nick away at me was that I wanted to contribute to this map. I wanted to actually be able to get in there and edit stuff.

But it was difficult, much more difficult than I thought it should be. I was used to Adobe Illustrator, which I used to draw traditional paper-based maps. In Illustrator, when you're drawing a line, you just go "click, click, click": you do all of the vertices one by one and it draws the line between them. The editing tools we had for OpenStreetMap back then were much more laborious, because they were just a wrapper around the data model. First of all, you did click, click, click, click to create your nodes; then you joined them together with segments; and then you joined the segments together with ways. Eventually, about 15 minutes later, you'd managed to map one road.

All of the editing tools back then were written in Java: the original Java applet which you and Tom Carden did, and then along came JOSM, this desktop-based Java editor. That was great for Linux users, but I was using a Mac. The Mac had Apple's own Java, which crashed all the time and worked really, really slowly. And so, whenever I tried to add a road to OpenStreetMap, it was no fun at all.

At British Waterways, I'd just rewritten the web maps for their website with this open source library called Ming, which let you build little Flash applets without having to buy £500 worth of software from Adobe. And so I thought, "Well, maybe I could use this to make editing OpenStreetMap as easy as Illustrator." That was 2006 when I started on that – and that was what became Potlatch.

You're bringing up all these sort of ancient pieces of history that I've forgotten about, like Ming.

Things I don't miss at all: installing Ming and all its dependencies. I wasted several afternoons trying to get that to install on the OSM dev servers.

It sounds like things that we do today that are fairly easy, used to be extremely difficult.

To a large extent, we were finding our way. To begin with we had nodes and then segments and then ways, and it only became apparent over the first three or four years or so that segments generally just complicated things. We could simplify the data module by taking those out. But we couldn't have stood there with any degree of confidence in 2004 and predicted, "They'll be a daft idea, we won't do that."

And the back end was also changing from XML-RPC to REST, and then got rewritten in Rails as well.

I remember that because I just spent the whole of one Saturday working out what this SOAP thing was and how I could talk to it. And then I think I sent you an e-mail saying, "Hey, I've just cracked SOAP." And you said, "Yeah, don't do that. We're moving to Rest tomorrow."

I was attempting to save you time.

Which of course it did! Again, I'd stress that I was really absolutely not a programmer in the classical sense at all. All of this was just stuff that I improvised. The initial version of Potlatch didn't actually have any concept of talking to a formal API, whether it was SOAP or REST or whatever: it just talked directly to the database, which gave people various palpitations. It took a while before Potlatch actually started talking in proper Rails objects to the website like everything else.

So it was a moving target. Again, segments are a good example of how the design changed. I decided for user interface reasons that I wasn't going to expose segments in Potlatch: it would let you draw nodes and they would get connected automatically. And I think that that was the final straw for segments: when Potlatch made it that easy to draw ways without going through this intermediate step, the user interface actually dictated what the data model should be.

We're on API 0.6 now. Every change has been some degree of heartache and some degree of trauma for those of us who write editors to cope with it. A lot of the reason why we've been on 0.6 for long is because so much is built on it. If you were to say, "We're going to move to 0.7 tomorrow, all these things are going to change," then people who write editors, people who write renderers, would say, "Oh, that's my weekend, my next month gone." Back then, we didn't have to worry so much about that. We could move faster then.

Yeah, I miss it. It drives me batty that we can't do that.

I know what you mean, but actually I'm starting to think that there are a lot of things that we can still change, we just haven't identified them yet. The website and API work great: they don't need changing much. There are big challenges in making OpenStreetMap easier to edit and easier to use, but I don't think an unchanging API is the obstacle there.

OpenStreetMap is very decoupled.

Yes. The advantage of OpenStreetMap being all these little discrete chunks is that you don't need to ask anyone's permission. Because there's an open API, anyone can come along and write an editor. Anyone can come along and write a renderer. Because it's not just one great big monolith, then improvements can happen in isolation without getting anyone's permission.

And sometimes they just come from nowhere. The best example of that was Mapnik. We had struggled for absolutely years in trying to get the map to look nice. The very first map actually pulled satellite images directly from the Landsat server, making an HTTP request every single time, and then drawing white lines over it – with Processing in Java, or something like that? It stayed like that for ages and just kept falling down as we grew.

So if you look at the mailing lists back in 2005 and 2006, the one thing that comes up most frequently is how to get an actual map on the front page that people can look at. And all of a sudden, this guy called Artem pops up and says, "Hey, I've written some software. It's called Mapnik. It draws maps. You should have a look at it." Holy crap, where did this come from? No one had any idea this was happening. No one has any idea he was writing this. And all of a sudden there's this beautiful tool where you can feed in all our data and make a lovely map with it.

So then you and Steve Chilton and Artem came around to my house, and we projected a laptop up onto my living room wall. And we started developing a Mapnik stylesheet, saying, "Ok, let's have these roads in this color and these roads in this color." Those, I think, are still the colors that are used in OpenStreetMap today – which is a bit worrying because thinking back, my living room at the time was actually painted yellow, so the OpenStreetMap road colors probably only make any sense if you're looking at a monitor that's got a yellow tint. If you're looking at it on a beautiful clean white dry monitor then the colors might not work so well. But hey, they've lasted.

I think that's a great example of where the whole is more than a sum of its parts. My memory is that you've got this sort of the Hot Ninja C programmer, Artem, writing this mapping software but has no idea how to do cartography; and then you've got Richard who can do all the cartography but not edit these crazy XML files. If you put the two of them together, then suddenly you get color maps.

Absolutely. And those maps were one of the things that really started to drive adoption of OpenStreetMap. The idea that suddenly it's a proper map, not just a bunch of hobbyist work: here is this map that looks as good as anything else that's out there, but you can go along and edit it.

And then of course you can add another component to that which is that we got some really good sysadmins: Tom Hughes, Grant Slater, Jon Burgess. They took what Artem had done with Mapnik and made it so that it would refresh pretty much instantly, so that you made your change and then five minutes later it would show up on the map. That was a huge jump for us. Initially we had sporadic map updates; that then became weekly updates on a Wednesday; but when we went from updating a map once a week to updating it pretty much in real time, it really started to take off.

Yeah, I think so. I wrote the original Ruby daemon that marked tiles as dirty and then put them in the rendering queue. It was quite simple, but then Jon Burgess who made it all work as a magical Apache C plug-in or something—mod

mod_tile.

— yeah, and then Matt Amos did the stuff to magically mark tiles as dirty when editing happened within the tile.

Yes – all of these different people come along and bring their own sphere of expertise to make something wonderful.

Were there any other open source projects you were tracking at the time? Wikipedia would be an obvious one. But I'm just curious what else was happening in parallel.

Not really. OpenStreetMap was my first exposure to open source. I didn't come to this from an open source background; I got my first Mac in 1992, I've grown up with Macs since then, and the whole Linux toolset was completely new to meme— the idea that you download some C++ and compile it.

Let's dig into Potlatch. I think we covered the motivation. But just to be sure, it sounds like the motivation was purely just to make something that behaved like a human might think instead of how a computer might think.

Exactly that. Someone who wasn't coming from a computer science background should be able to come along and just add their local knowledge. We're not interested in your computer skills; you know something about your location, and that's what we're trying to gather together.

And so, the tooling at the time – the obvious thing to use was Flash, which had started as a way to sort of try to make the web a little bit interactive. But it wasn't quite a real programming language.

Flash back then was ActionScript 1, which is like JavaScript but without the whole ecosystem of "this is the proper way to do it". Flash was mostly marketed to designers as "here's a way of animating your little stick figure to walk across the street", or more often, "here's a way of getting your banner ads to flash in various ways."

So it was a reasonably primitive programming environment. That suited me fine because I was a reasonably primitive programmer. But there were certain limitations as to what it could do: it couldn't speak XML to the API in the way that anything else would, so you had to go through this Adobe-only binary format called AMF. So some of the design decisions for Potlatch were pretty much dictated by the way that Flash worked.

One of the big differences with Potlatch 2 is that Adobe moved on to a new dialect, ActionScript 3. This had a whole framework around it, Flex, which gave you lots of user interface components – tables, tabs, dialogue boxes, all of that. ActionScript 1 didn't have any of that. With Potlatch 1 everything was homegrown: all the dropdown menus, all the buttons, all the dialogues were code that I'd written saying, "If you want to do a dialogue, draw a gray box in this particular place on the screen and draw another grey box and put the word 'Okay' on it". Which started to become a problem when OpenStreetMap took off outside Britain, because I'd designed all these grey boxes to fit English language text. And so when someone comes along and wants to translate it into Icelandic, all the text runs off the side of the grey box and it doesn't look so good any more...

So you started writing this editor. Then what?

Then it sat there. It sat there for a few months because I didn't know how to integrate it into the website at all. The code was quite eccentric in its various ways, and I think you and other people were a bit—rightly—leery about it.

So we had a hack day in Oxford, in the St Aldates church hall, when the port of OpenStreetMap to Rails was meant to finally happen. You'd mostly completed the Rails code by then: I was going to sit down and try and integrate Potlatch into it. But I started looking at this – having never seen an object/relational mapper or anything like that before – and it might as well have been in Swahili. I sat down with my head in my hands thinking, "How does any of this work?" I eventually managed to bodge something together that was still largely SQL but nonetheless used the Rails way of talking SQL to the database. I called you over and said, "Look, this isn't how you're meant to do it, but it's working". And I think at that point you said, "Oh yeah, that'll do." And so, Potlatch went live pretty much the same time as the Rails port did.

Richard Fairhurst

Do you want to talk about tagging?

The freeform tagging approach is what has made OpenStreetMap so rich: you can come along and map whatever you like. We have certain overarching principles. You write down stuff that's factual rather than stuff you've just made up, and you try and coordinate what you're doing with other people. But that's pretty much it.

The story of OpenStreetMap tagging is the story of putting a community together and letting them work out things for themselves. Andy Robinson drew up the original tags all those years ago, but since then there's never been any oversight, any direction: it's just happened. Sometimes, it's been influenced by the presets that editor authors have chosen: I can say, "Right, I'm going to use these particular presets in Potlatch," and that gives a tag a head start because a lot of people just going to use what they're presented with. But it's still evolved in a million and one crazy directions. People now are using the tagging system to express 3D geometries – stuff it was never designed for, but it works. Just about.

At first the freeform approach wasn't something everyone was happy with. "We must have structure to this. We must do it according to this external way of doing things. Or we must have everything organized into a strict hierarchy." But none of the ontology people can ever agree amongst themselves. So you have us all doing it freeform and then you have someone who comes and says, "No, you should use this standard." Then someone comes along and says, "No, you should use that standard." Meanwhile, while the standards guys are arguing with each other, we just get on and do it. And eventually it got to the stage where the freeform method had such momentum that you couldn't change it even if you wanted to.

In the first five years of the project, how did the community work?

To begin with, the community was very dispersed, very international. You had people like you in London. You had people like Lars – you remember him?

I do.

I think he was in Norway. You had a couple of people in the States. And it really was mostly an email list plus a couple of people who got together in London every now and then. As the project grew, you started to get a more of a concentration in particular places, and people could start meeting up. But to begin with, it was very much a remote experience.

The first time that I met the core of what's now become the London OSM community was at an event you organised called Techa Kucha. Speakers gave five-minute talks, and we all went to the pub afterwards. That was the first time I met Andy Allan, who did so much for Potlatch 2, and a bunch of people who are still playing a crucial part in OSM today.

Over the years, a lot of the most interesting connections in OpenStreetMap have been made in a pub like this – I remember, after working out the cartography with Artem and Steve Chilton, we all went to a pub in Oxford and I think I had a standup argument with you about licensing! But this is how the connections are made. You have a relaxed, lubricated environment where you can bounce ideas off of each other, and say "Yeah, that might work" or "No, that's a lot of crap," in a friendly way. Of the design decisions made about OpenStreetMap over the years, then a pretty high percentage will have come from a meeting in a pub.

Thank you for making such a neat segue to licensing.

As soon as you've got a load of open source people together in one room or one mailing list, they are going to argue about licensing. That's nothing specific to OpenStreetMap or open data: it happens. It's one of the fundamentals of open source that people enjoy arguing about licensing. It's quasi-religious… you can't convince people. Which doesn't stop you from trying, of course.

Back when OpenStreetMap started, you chose CC-BY-SA, the Creative Commons Attribution-ShareAlike license. I don't know to what extent that was a "I have thought about it for two weeks" decision or "I have thought about it for five minutes," but whatever you'd chosen, it would've been wrong for some people.

Because we were the first massive collaborative open data project, no one had thought through all the licensing implications before. So there was a lot to work out and a lot to discuss. For some reason, I found some of this stuff quite interesting and started to read up on some of the issues around it, on some of the challenges.

And that started this six or seven-year conversation that eventually led to changing the license into something that works best for us. But of course, I say "it works better for us" but the same still applies: you're never going to satisfy everyone with the license, no matter what you choose. There are still people saying, "Yeah, but it should all be public domain" or whatever, and that's always going to happen.

Do you want to describe, in terms a five year old can understand, the spectrum of points of view on the license?

The big religious debate in licensing is between permissive and reciprocal. Permissive means you can do whatever you like with the code or data. The best example of that is actually the license that I've applied to Potlatch – the WTFPL, which stands for the What The F*** Public License; you can do what the f*** you like with the code.

The other end of the spectrum is reciprocal licensing that says, "You can do whatever you like with this code as long as it retains the same license" – so it has to stay open. You can't take the open source code, or data, and use it in your own proprietary product.

Because a lot of early OpenStreetMap people came from an open source software background, they arrived with their own pre-conceived ideas. They had seen all this before, so some were already permissive advocates, some were already reciprocal advocates.

And so the conflagration starts with a fair amount of fuel on it already. But no one had really thought through, "Okay, what does reciprocal actually *mean* for a data project? In what circumstances do you have to give back? If you make a lovely printed map which has this beautiful choices of colors on it, or these beautifully drawn symbols, do the symbols and the colors have to go under the same license or can you keep those as your own intellectual property?"

The same goes for software that uses OpenStreetMap data – for example, routing software. Does the routing algorithm also have to go under the same license?

All these things we had to think through for the first time. There were various degrees of shouting and accusations, in the way that mailing lists arguments tend to go. But actually there was a fair amount of clarity, because mailing list and pub arguments do at least force you to think things through properly. If there's a single hole in your argument then someone is going to come along and say, "No, you're completely wrong. You're an idiot and here is why you're an idiot."

That's quite a way to end a sentence. What was the role of the OpenStreetMap Foundation in this?

The OSMF was born because the assets – the domain names, the project name and the servers – have to belong to an entity with some sort of constitution, rather than to a person or to a bunch of people. And what falls out of that is that if the foundation controls and owns the servers and the domain name, then it is the body that ultimately is in control of what is hosted on those servers at that domain name.

The individual contributors in OpenStreetMap had always owned the rights in their own contributions: you have the rights in all the mapping you've done. But the license for the content hosted at openstreetmap.org is the Foundation's decision.

So first of all, we had to come to some sort of consensus as to what an ideal license for OpenStreetMap might be. When we'd done that, the foundation had to agree that yes, we want to host the database on our servers under that license. Then we had to get the agreement of everyone who had contributed to that database: "Yes, I agree to change the license of my contributions from the old one to the new one." And then finally, we had all the hard work of stripping out the contributions made by people who hadn't agreed to this, so that people could download the new OpenStreetMap database with confidence that it was available under this license.

The arguments went on so long that we only got to the stage of "Let's ask people to agree to the new license" really late in the day. And by that time we had hundreds of thousands of contributors. So how do you get hundreds and thousands of people to agree? We did and that's the biggest miracle of the whole thing. I don't know any project that has managed to pull that off, apart from us. Look at Wikipedia. They changed their license through a hack: they basically rewrote their existing license to say "This license automatically gives the right to change it to another license. By the way, you've all agreed to this." Imagine trying to get that through the OpenStreetMap mailing lists – I think you would be able to see the explosion from several continents away. No, we had to ask everyone, and we did.

There were a bunch of people who were involved in the license change: drafting the many revisions of what we wanted from the license, working with the lawyers who drew up the license, doing the technical work to actually get the database migrated over. But really, the main work was trench warfare on the mailing lists, every single day, until you got a consensus that this is the license that will work best for us.

We didn't convince everyone. There were communities who were not having any of it—in Australia, in particular, for social reasons and for particular circumstances to do with Australian government data. But overall, we got it through. Of course, some of it may have just been that people thought, "for goodness' sake, this has been going on for eight years. F*** it. I agree."

There were people who were convinced that the change was being made for nefarious reasons. And it genuinely wasn't. Seriously, if any of us were doing it for nefarious reasons, we'd have done a much better job of it. We would be much better at being evil. But it wasn't. You didn't, all of a sudden, see that two days after the license changed, you or I or anyone was able to sell out for a billion pounds or say "Haha, we won!" No, that didn't happen. It couldn't happen.

My side in the religious argument is that I like permissive licensing. I like the sort of license that says, "Do whatever you like with this." But it's not about what I like, or what you like, or what anyone likes. It's about what will get through the community, what the community will agree to. It became obvious really, really early on such even if there were a majority for permissive licensing – and maybe there is – we weren't going to get it through. There was a sufficient weight of people who like the reciprocal licensing that we've always had. Again, today, there are people agitating for a move to permissive licensing. I can see their arguments, why it would work better for them and in some ways for OpenStreetMap in general. But the community balance right now is not there and I don't think it's going to be.

If I were 10 or 20 years younger and had my time all over again, I'd love to do a PhD in the dynamics of a community like OpenStreetMap. Because the community is fantastic. The community is what has built this project. It's very easy to laugh and to sneer at all of our eccentricities – and with good reason a lot of the time – but nonetheless the map has only happened because of the community.

We've only got where we are because we have really dedicated contributors – you could say obsessive. But when, in 2006, we wanted to make maps of housing estates, without any aerial imagery, all we could do was trudge around with little Garmin GPS sets, writing down all the street names. You've got to have obsessive people for that, because sane people wouldn't do it. And so, it's due to all of this that we have OpenStreetMap as it is today.

But that poses its own challenges. If you put a bunch of obsessive people on a mailing list, they start fighting. And when the project grows big entirely due to the dedication and hard work of these people, then you've got to work out how to make it appealing to people who've still got stuff to contribute, but perhaps don't want to invest as much of their time into it as the other guys did. And that's a tension that we're still struggling with today.

Still, we've evolved and improvised really good ways of working together. I would love to do a proper study of that and say, "What has worked? What hasn't? And how did we get there?"

So do you want to talk a bit about Potlatch 2?

Potlatch 2 was essentially Potlatch 1 done properly. It wasn't written entirely by me, which was a good start; we had more people chipping in on it. It was written in a more modern programming environment, ActionScript 3. It's a properly written application with a lot more polish, a lot more colour, and in many ways was what we needed to take OpenStreetMap onto the next stage in terms of contributors.

So rather than just having this fairly basic display of colored lines, all of a sudden, you have something which looked like a map, which had the road names written on it, which had nice colorful icons and so on. It had more tools. It had a nice easy user interface to abstract away all the details of tagging, so you didn't have to remember that it's "highway=primary" rather than "highway=main". You could just go to a dropdown and choose the picture that looked like the road.

In particular, we had Dave Stubbs who was actually a proper programmer and knew about proper programming. He came along and said, "Right, if you're going to write an editor in ActionScript 3, here is how you should do the architecture." Brilliant, thank you very much, because I wouldn't have known where to start. That gave us something that was robust for the next few years.

So it became a code base that people could understand, and that people could make contributions to without having to wade through 2000 lines of my spaghetti code. I remember we had one hack day here in my house in Charlbury, and we asked "What can we do to improve Potlatch 2? What can we do to make it better and more useful?" And Matt Amos said, "Well, you know, the thing I always find myself using in JOSM is the quadrilateralize function" – the tool that makes nice square, right angles. Matt understands maths, which I don't, so he was able to write this pretty quickly. Brilliant—that would have taken me absolutely weeks to do.

Similarly, a pull request came out of nowhere with code for multiple selection, so you could select five items and change the tags all at once. Again, I would never have even known where to start with that, but someone just wrote 200 lines of code and it worked.

The editor experience has been iterative. Every two to three years, we get a new editor that improves on the last one. We started with the Java applet, then we had Potlatch 1, then Potlatch 2, and now we have iD. No doubt, in another few years, we'll have something new using the technology of the time, designed for the sort of contributors we're trying to attract. And, you know, that's just a matter of the project growing up, the project surviving. You have to keep reinventing yourself like that.

The challenge is converting people who are interested into people who can actually edit. So when they go to openstreetmap.org they are able to click something and correct the road name, rather than going to it and saying, "I don't understand any of this."

The editors have got more user-friendly as the project has become more mainstream. Back in 2004, it didn't matter that we had a rudimentary editor because the guys who were interested in the project were by definition computer geeks and could cope with a rudimentary editor. Right now, OpenStreetMap is big news and big business, it powers all of these consumer-level mapping sites, so we have to have something which is much more accessible.

And that's why we have iD which takes everything on another step in terms of user friendliness. Potlatch 1 and Potlatch 2 were both steps along that way. You can see ideas evolving across the three, that the stuff that Potlatch 2 takes from Potlatch 1, the stuff that iD takes from Potlatch 2, and I'm sure whatever we have next will take stuff from iD.

We give the editors project names. We call them Potlatch. We call them iD. But they're not, they're new versions of our editing app really. To begin with, Potlatch didn't have a name: I just wanted to call it "the edit tab". But it turns out that in open source, you have to give everything a project name so I gave it this daft project name that no one could spell. When I started thinking about iD, I thought,

"Well, this is ridiculous, no one can spell Potlatch so I'm going to have something that no one can fail to spell. I-D. Two letters." Of course, what I didn't realize is that everyone can spell iD but no one can Google it.

iD is a really nice example of how development has moved on. It became obvious that Flash was not long for this world. And it also became obvious that we needed another step change in ease of use. I'd come to the conclusion that the user interface could be a bit more modal. Potlatch has never had modes, no toolbar or anything like that: your editing behavior depends on how you click, how you drag. You don't have to select different tools like you do in JOSM.

But when I started on ID, I thought, "Well, maybe we can make the user interface more obvious by having a few really simple modes: one for add point, one for add area." I got some way with an early prototype of ID in JavaScript using this slightly corporate framework called Dojo, which in some ways was similar to ActionScript 3 so felt natural to me. I stood up at the State of the Map US conference and said, "This is where I think we should be going from here."

What was really nice was that then other people, in particular John Firebaugh and Tom MacWright, said, "Cool. Let's build that." The end result actually has virtually none of my original code in it. It's come on a long way since then. But it's written by people who really know what they're doing in terms of programming and in terms of user interface. It will serve us very well for the next couple of years, I've got no doubt.

What's your involvement in OpenStreetMap today?

I'm a mapper and a contributor as I've always been. But I'm also using OSM data as a consumer for the first time. I've set up a website, cycle.travel, which is bike routing and cartography based on OpenStreetMap data: an attempt to build a really good quality route planner for cyclists based on OSM data.

And, as ever, I continue to look around for the next thing. For me, the really interesting challenge is mobile editing. At the moment, we're doing really well with desktop editing. But for the past few years, everybody has been walking around with these amazing computers in their pockets with cameras and GPS receivers. They should be able to contribute to OpenStreetMap really well using that technology, and we haven't cracked that at all. We've got a bunch of decent mobile editors but they are as effectively recreating the desktop experience on a mobile.

Let's say we want to get addresses for every house in the world. Right now, using the existing editors, you have to note down all the house names, plot the nodes, enter the tags and so on and so forth. I should just be able to walk along the road with my smartphone and dictate into it "Number 15, on the left. Number 17, on the left. Number 19, on the left." By the time I've got to the end of the road, it's mapped. That's a huge opportunity for us and we haven't got there—yet!

Ben Gimpert

Bio

Ben Gimpert is a researcher in machine learning and systematic trading at BlackRock in San Francisco and London. He has a long-time commitment to performance, beginning as a US national champion roller skater (1998 USARS Nationals, Fresno, CA); hosting promenade-theatre-style alternative reality games (e.g. Requiem, Pleasant Home, Oak Park, IL 2000); and later training as a classical French chef (Le Cordon Bleu, CHIC, AASc 2004). Before moving to the Left Coast, he worked for Encyclopaedia Britannica (Chicago), JPMorgan (London), Credit Suisse (London), and ThoughtWorks (Chicago). He was an early contributor to the activist, grassroots cartography project OpenStreetMap (since 2005), and was online before the Internet, as a BBS scenester. He coded his first visualization project as a "demo" in the early nineties, in Turbo Pascal. He has an MSc in Finance from London Business School and a BEng in Computer Science from Northwestern University. He dropped out of his PhD programme at The Bartlett Faculty of the Built Environment, University College London.

Favorite Map?

"I would probably choose a TIGER-derived map of Manhattan, since that was the first thing I ever really focused on in the project."

Interview

Why don't you talk about who Ben was before finding OpenStreetMap?

I had moved from Chicago to London in '04 to go to grad school. And also, I was chasing a girl. I had always wanted to live in Europe. I wasn't necessarily certain I was going to end up the U.K. but that's how everything shook out. Then at grad school, I was at University College London (UCL) and that's where I met you, Steve. I'm trying to think of the actual introductory moment. It was probably Alastair, right? Rest in peace.

I'm not sure.

I'm trying to remember. I mean, I can remember very clearly going to the pub on Charlotte Street right next to UCL and actually getting to know each other and having some pints. I'm trying to remember who was the actual impetus. It might have been Alan Penn or Alastair Turner that actually said, "You guys should meet each other." Because I was technically a part-time PhD student, so I was not around the Bartlett every day and I think you were kind of fluttering in and out, doing different amazing projects. And so, I don't know what the actual impetus was but you and I hit it off. And so, even though OpenStreetMap has grown into being kind of its own ecosystem, it was you and your personality that actually kind of turned me around because up until that point I had not really been like even a shitty cartographer. I mean, I'd just been pretty much unaware of the GIS world. And so, I joined the mailing list and started lurking for a while and then helping out later.

And what were your thoughts when you first saw it?

I thought the premise was completely—you know, it's such an important project that when you first experience it, it seems obvious, right? Clearly, map data in a crowdsourced form makes sense because it's so driven by experience. People walk around the world, observe space and then have technology like GPS trackers and breadcrumbs and whatnot that they can provide to a database. So, it seems obvious and clear from the start.

The part of it that was a little bit more specific to me is that—was I the only non-Brit involved in OpenStreetMap at that time? Maybe, I might have been the only one. And being specifically American, I was spoiled by how much census data was open and available online. So, even from the very start, watching the mailing list, all these passionate people that saw the Ordnance Survey as the justified bogeyman—not really a bogeyman, but the Ordnance Survey as the gorilla in the room for proprietary data. I guess, I couldn't relate as well as folks who had struggled with proprietary map data for ages. So the long answer to an easy question is, I started out pretty early on thinking ahead to OpenStreetMap version two. So, once we had a good crowdsourced map of the U.K., and then The States, and the world, then the metadata and the buildings and commentary and all of the geographic sort of neighborhood-based data that would end up being the next layer of Open Street Map—that was the part I found most interesting but we didn't get there for a little while.

What was interesting about it?

I thought it was more specific. I figured, at some point, I would end up back in The States where I'm from. And so, from my selfish standpoint, it was what's going to be most relevant to me because census-derived U.S. geographic data was good enough. And so, I felt like OpenStreetMap's real place to shine for Americans was that metadata. So, in a real practical way, it was the part that I found most relevant to me. But, I guess more philosophically, I think the metadata is kind of the richer, maybe more editorially-driven spatial information that people are going to want.

Right from the beginning, I was thinking about pub reviews. I'm not sure if he still runs it, but my buddy, Arthur Rabatin, at the time ran Fancyapint in the U.K. I always imagined that stuff as a layer on OSM once it's done right. And then you've got routing decisions that could be made not just by efficiency of time or gasoline but more like "take me home via my favorite donut shop" or whatever.

Do you want to talk about the first hack day?

Ben Gimpert

That, I believe, was at the Bartlett and we sort of, you know, it's a while ago now. It's getting on ten years so I have a hard time in my mind differentiating the first hack day at UCL from also the one that was at Limehouse Town Hall.

The time when I sat down and banged out a little bad Ruby code to tease out some TIGER data, and eventually it was going to end up imported into OSM. I think that was the goal early on. I think that happened at Limehouse which is that kooky old space in London where they also hosted the Dorkbot events. And so, I think it was like a Geo meet-up. Of course, ten years later, I'm thinking that in my mind I might be remembering the hack day as that too.

Would you want to talk in generalities about both of them?

I remember, pretty early on, there was a schism between the very Nathan Barley-esque "talkers" and people who could code. I mean, it's my own engineer's snottiness coming out. Even being able to write some shitty SQL is better than someone who has grand ideas but is not really able to make any of them come to fruition. And I think that's where people like you and I met Matt Amos and Tom Carden and kind of clicked. We were all practitioners, even if we were not expert developers or GIS people from the start. I was a good coder then but that's been my sort of lifeblood forever.

In this project that was going to go on to have huge intellectual property questions and had a provocative, activist angle from the start with the Ordnance Survey as the bad guy, I also saw that there was a pretty early schism in the Geo community in London between makers and talkers. But from a real practical sense, I remember just huddling over a laptop and banging out some code while someone was talking in the background upstairs at Limehouse.

How about the first anniversary?

I remember the first anniversary was with you and I. I think Kate Elswit and Alex Lotinga were there. Tom, Amos—also, was Nick Whitelegg there or Ian?

No. I mean, I only remember you, me, Tom Carden, Kate and Alex.

I think we were at that pub in Lower Camden Town. It was called The White Lion, which narrows it down to about a thousand pubs in the U.K. but I think it was where we were on the picnic bench outside and the weather was actually decent.

My memory is, it was getting dark and we were inside. The primary thing is that it was small.

Oh yeah, it was tiny. Oh yeah. I remember, at most, it was you and a couple of people and then dragged along partners and basically bitching about the Ordnance Survey. Although I do remember another event. I don't know if it's necessarily the anniversary but I remember it being pretty formative. We were sitting at a picnic table. The partners were dragged along as well. I think Nick Whitelegg was there. Amos, you and I, maybe even Nick Black. And I think it was really early. It might have been like the second year or the third year of the project. Do you remember this one? I remember pitchers of beer which was not necessarily typical for the U.K.

Ben Gimpert

This is why I'm doing this. It's to capture these memories, you know?

Absolutely. I remember that one as particularly vitriolic politically. One of the things that strikes me about early-on in OpenStreetMap's history, and we love it and hate it in hindsight, is how much University College London (UCL) and The Bartlett were the default physical spaces and the servers and bandwidth. As time goes on, I kind of think of them as having sponsored the project even though it was unofficial. It was always us plugging Ethernet cables in and breaking the Wi-Fi, but the fact that it was all very UCL is interesting. This was not a project that came out of academia, but it was nestled-in and cozy there.

Do you have any memories of Dorkbot that you want to talk about?

Dorkbot is still going on. It started in New York twenty-five years ago. The London events were particularly diverse. I remember it was a lot of folks doing weird synthesized music and GIS stuff like us, and also even cooking and performance art. They have some slogan where it was "Tech people who love artists, artists who love tech people, and those who don't care about the difference."

It was a monthly gathering, maybe you had to pay a bit of money or were encouraged to buy a beer. They'd organize speakers. The best thing about Dorkbot in London, other than all the characters that would come out of the woodwork, was that I remember it being in Limehouse Town Hall which is a particularly London space, this dilapidated, falling over, community structure. Nothing would remotely pass code. It was in a liminal neighborhood, and the place was just full of junk and signs.

I remember Saul Albert being the host several times. At one point, I presented another project of mine, a flavor engine for food and cooking called Holy Shallot. It was an engine for automatically suggesting flavor pairings and designing menus. And another time, it was OpenStreetMap of course.

I remember Dorkbot being very charged-up and fun. It was a lot of talkers as well as makers, but it was unapologetically artsy, which I liked, as well as being very tacky. It was a monthly get together of about a hundred people listening to other folks give talks and show demos of weird stuff. Everything from robotic dildos to synthesized music and weirder projects.

So, jumping back to TIGER, do you want to talk about what TIGER was and is and what you did with it and why?

TIGER is from the census done every ten years in the States. One of the outputs of the census is a decent dataset of streets, roads and classic GIS atoms. So geographic features. Because of cultural norms and just history and legacy, this dataset is made public and available. I think it's just public domain in the U.S. So if you're building a GIS tool in the States, it tends to be your first messy dataset.

When OpenStreetMap was just starting, one of our thoughts—my thought, your thought—was to get U.S. TIGER data imported into OSM early because you might as well. It's sitting there. Though the community in the early days was focused on the U.K. and the Ordnance Survey bad guy, we thought that TIGER data would be a good test of the system. It also spoke to my desire to skip ahead to the metadata.

From a practical standpoint, it meant me writing some Ruby and pulling in the actual TIGER data, figuring out this kind of old school columnar schema, and throwing stuff into the OSM database. At the time there was no web service, no API as such. We were inserting rows. That was before we had migrated over from MySQL to Postgres or something like that. I can't remember the details, but it was at the point when there was going to be a transition in database infrastructure too.

I started with doing Manhattan because I had family in New York, and I thought Manhattan would be a very recognizable, cool image to have as part of the OpenStreetMap. From a speed perspective, it was tragic. I remember there were some serious resource constraints, so the TIGER import process was this thing I babysat in the background for a little while and it would always fall over and pick up some line type that didn't work.

Yeah, my memory is that we figured out some months later that it was just re-importing the same stuff on top of each other multiple times. And eventually someone did a re-import.

I think it was probably that re-importing process that was tapping the database.

That's great. I mean it speaks to the learning process. Do you want to talk about how both the project, the API, the data structure, everything was a moving target?

You use those terms now and you should, given how the project has matured, but at the time I'm not sure any of that was applicable. It was a nice ambition in our minds, but there was no API. We were mucking around with the databases. The great metadata tagging wars were bubbling in the background of the mailing list, and they would flare up once in a while. But in the early days, it was really a focus on getting the data side of things and the queries functional.

I remember you and I going back and forth trying to figure out good ways to do joins and do the real grunty DBA to get the queries even remotely feasible. Someone early on had the idea of doing what we would nowadays call a quadtree. I think Matt had the idea to do it using the Subversion protocol, and we would have a quadtree that was a nested directory structure within the source control space. We liked this idea because it was implicitly versioned and historied and persistent. Nothing ever came of it. I think we did better with just thinking about the database, but it was very fluid.

Yeah. I think his idea was actually tetrahedral.

That's right.

Cutting the world into triangles. I don't remember about Subversion but that makes sense. My memory is that that sounded really great but who was going to go build it whereas just having two columns in a database for latitude and longitude seemed like we would achieve something.

The GIS library support was really crude, right?

It was MySQL. It was MySQL in the beginning and I think it was sort of a chicken and egg, because the software and tooling was pretty crude because there was no data. Which meant it was hard to build things and go collect data.

My favorite memory of OSM was actually at some point we were taking a black cab on from a pub to a party. This might have been the first anniversary party, or in my memory it is. In the cab, I remember you pulling out a GPS tracker device that would keep a digital breadcrumb of where we are moving in the cab and throwing it on the roof of the cab with your arm out the window. And this was one of those nights when partners were in tow, and where the activist side of things felt very like Hollywood.

What do you mean?

Trying to explain to a black cabbie why the hell we were driving down the street with our arm and a GPS tracker out the window.

Well, what do you mean by Hollywood?

The action scene of it. We're talking ten years ago, so it was not a default that everybody had a variety of cute little gadgets in their pockets that do GPS tracking. I mean, to pull out a bit of kit from your pocket, tell a cabbie something about a government organization that keeps all of this data locked up and proprietary, and that we're going to fix it. But part of it is we're going to keep a breadcrumb of this cab's journey through London and upload it to a database later.

Yeah, it all felt very natural at the time but as an outsider—you know, if an alien landed and wondered, "What are you doing?" Which part of this is rational?

Which part is rational and wow, this bad guy does exist! It's not just tin-hat paranoia at the time.

Did you go to any OpenStreetMap conferences or the Isle of Wight?

The Isle of Wight: In the history of the project, I do not think it was momentous but it sticks out in my memory because I got to know a lot of people. This was putting faces with names. So that's where I got to know Nick Black and his partner at the time, and Etienne.

Why don't you tell that story?

I remember his extreme paranoia.

Yeah, so tell the story from the beginning, assuming people don't know anything about it, or crazy people or OSM.

You ended up writing a note, putting something in writing. I don't know if it was for funding or some partnership that involved Etienne. It was something you wanted to have a record of, and I gave you some advice. In The States, one of the things you can do to establish a very legal time stamp on some bit of information is that you write it down and send yourself the document in certified mail and don't open it. A government puts a timestamp on something physical.

I think, my memory is, he was so weird and we just wanted to write down, "The weird guy showed up, called himself Etienne." Maybe we're on some secret list somewhere, you know?

The Isle of Wight Conference was cool because it's a part of the U.K. I hadn't been to before. It was nestled away. It was a combination of a mapping party and also just putting faces with names that I had known from the mailing list. There were some people from the continent who showed.

There were two groups. There were the folks who spent a lot of time literally mapping the Isle of Wight. So running around with GPS breadcrumbs and getting this beautiful, somewhat obscure island off the south coast into OSM. There was also a group of us who wanted to use the excuse of face time to do some technical work.

I remember holing up and thinking about architecture, infrastructure, and the political side. There was this question of how much fun it was to map. Some of us loved it, some of us found it pretty boring to run around gathering the GPS breadcrumbs.

You might want to edit this part out but I remember you saying, "I don't really like this. I don't like running around mapping too much" even though it's super important to the project.

Do you remember the pub that we all went to with the maze at the back?

That's a flashback, hell yeah. I remember Nick Black and his partner, I think we crashed in the same cabin.

That was like the first time you and I talked about "how sustainable are mapping parties as a way to map the world?" because it appeals to a relatively certain type of personality. This "love of space, I want to record it and breadcrumb it for the world" versus "partnering and figuring out larger scale ways to import data, tracing imagery, or asking big companies to donate."

Do you want to talk about your Etienne story because my memory is that was towards the end, it was like the end of Sunday or something like that?

Well I remember him literally not wanting to show his face. And you and I thinking he was either bonkers or secretly a billionaire. We weren't even sure about his name, whether that was truly his name.

That's what we're missing from the transcript is the beginning, like what happened? My memory is that towards the end of Sunday, there was a guy who wanted to come out and do some mapping and he showed up but wouldn't tell us his real name. And the conversation was very weird and obscure to the point that we weren't sure if we were on some sort of secret list or something that they had sent

someone to find out if we were anarcho-terrorist mappers or something.

I remember being outside on Sunday. It was a little chilly and we were on a lagoon or a canal, by this garish primary school-colored railing. Etienne showed up on his bike which threw me off a little bit given that it was an island. Maybe the government did drug us, because it was pretty stunning but I only remember the after effects. I'd tucked away the note but didn't do the certified post thing. In hindsight, it didn't seem too important.

Do you want to talk about, you know, over this time period, what you're working on and how that was changing?

I started out in grad school at UCL, the de facto early OSM sponsor—UCL and The Bartlett. I was a PhD student working on traffic flow, Space Syntax, how people move in patterns or lack thereof on a floor plan. My ambition had been to parlay this into a view on virtual worlds, and use some of the stuff I'd done as an undergrad that was an early Minecraft.

I was doing my PhD part time, and I spent about half the week at a job where I was on a trading desk at an energy commodities firm doing modeling and software development on pricing and a risk engine. Though I found projects like OSM fucking cool, my actual research was going nowhere. My supervisor was pretty distracted, and I was distracted. I was also paying tuition through the nose. I didn't have too much funding help. Even though the U.K. has a reputation for being inexpensive for grad school, it was not inexpensive for me. I was surprised at how much I liked the quant work I was doing, I basically left UCL. I had a dormant PhD year where I was on the books, but eventually I started doing more and more quant work.

I was still involved with the OSM community but I started to pull away from it. I was tired of the perpetual licensing battle, and the Ordnance Survey as the bad guy felt less relevant to me as someone who was going to end up back in the States. So I got less involved with the project. I obviously stayed friends with you and Tom. I don't think there was a moment like a big schism or an argument. I think what did it for me was one of the very long, drawn out legal licensing battles on the mailing list. Eventually I unsubscribed. This was before legal split into its own mailing list. I unsubscribed and that was it for me. I transitioned away from OSM but then came back a bit when ZXV and CloudMade got into the picture a couple of years later.

And so, I dropped out of UCL PhD and started working full time as a finance & banking software developer and quant. And then eventually I went to London Business School and got a degree on the side. A couple of years later, I moved back to the States, to the Bay Area where I am now. I moved back in the fall of '09 to do something a little bit more entrepreneurial.

You, Tom and I all ended up moving from London to San Francisco.

That was one of the jokes my then wife and I had while we were still living in London is, "All the cool people have moved back to the States, why are we still here?" I mean, obviously, that's not true. I did and still have great friends in the area, but you had moved to the States. Tom Carden had moved to the States. There was a migration that happened, and I'm sure there's some cliché you hear about the pull of Silicon Valley or some shit. I don't know. I think a lot of it was also just a post- grad or post- undergrad transition.

The other thing I wanted to say is I think dropping out of the PhD and going into finance were probably the best financial decisions you've probably made in your life? I don't know about you but I don't know any rich PhD's in academia.

Yeah, that's the truth. And of course, I'm not sure I think of myself as rich but I mean, I don't hurt for it like a lot of folks do. Especially academics. I'm happy to admit when I sell out in certain ways, but I actually like the work. That was the thing that surprised me most. Stochastic calculus, quantitative analysis, derivatives pricing, no arbitrage pricing and the markets -- I found it fascinating. I loved it.

"Okay, so this is the evil banker stuff that everyone makes fun of? This is actually cool shit!" I remember I was on the trading desk at one point. This was when I was at J.P. Morgan. This was in 2008, so right as the credit crisis happened, I was working on a credit derivatives trading desk. It was an index products desk. I heard the trading floor buzz bump up one afternoon, and a bunch of chats were flying back and forth on Bloomberg. It was this natural like "Hey, what's going on?"

Through various informal methods, someone knew that a conflagration had happened in the Middle East. Someone shot a missile. It was one of the Israeli-Lebanese wars, I think. This had happened and it wasn't anywhere near the BBC news. It wasn't on Twitter yet. It wasn't anywhere yet. Yet, I felt pretty close to the heartbeat there. So it was the combination of the capital markets being very close to the news, and actually liking the math and the software side of things. I mean, I had no problem selling out. I'd do it again, any day. I even convinced friends to work with me.

So the hubbub rising when the market's changing reminds me of Neil Stephenson books because in his trilogy 'The Baroque Cycle' I just have memories of similar things happening in the exchanges in like the 1600s or whatever it was.

Yeah, the Tulip Exchange.

And of course, news traveled a lot slower then, right?

Oh yeah. I'm not sure it's the case now. I mean, I still work in finance to some degree but it's a more specific focus, the real estate world. I'm not on the heartbeat and capital markets anymore, unfortunately.

So have you been using OSM in your day job at all?

Definitely. They used it as a base map for some visualizations. They had a pretty map that's on a dashboard carousel in the office that was running non-stop, where you see where the company's focused at the time.

What's the company? Do you want to talk about who they are and what they do?

I joined as an individual contributor last summer. One of my startups had gotten acquired by a real estate firm so I'm starting to get to know residential real estate as an asset class. The company does a highly-expedited residential real estate sale. A homeowner comes to the site, and says they're interested in moving because they need to relocate for their job, or their home was something they inherited and they want to move very quickly, rather than the typical three months lag of hassle and paperwork that goes behind a residential sale. They expedite and do a lot of the bureaucratic paperwork behind the scenes and give folks an offer right away on their home.

They do a lot of GIS work internally. I described the visualizations that we use as just kind of the heartbeat of the company in the background, but we also use GIS features in our valuation and our risk management. So they look at things like "Does a home back onto a busy street? Is it on a lake? Is it on a golf course?"

That's kind of fascinating, those metrics. Like I'm assuming there might be things like how far is the nearest McDonald's? How far is the nearest grocery store? Those things must affect pricing and pretty interesting data analysis.

As a tech startup, that's really what they're trying to own. I mean, it's not a secret that proximity, walkability, noise, view -- all of these things help determine a home's value. How to capture that information as a feature for a machine learning or heuristic model, how to capture that as number that we can feed into a model and then get some confidence in whether the numbers coming out are valid, that's the tricky part. So sourcing the data is a lot of the work, and then actually getting it into a shape that feeds the model well is the other part.

It reminds me of a GIS joke. There was something that made the rounds on the Internet, in 2014 I think it was, like La Quinta is just Spanish for "next to Denny's." And then there was someone who actually went and found all the La Quinta's and all the Denny's in the U.S. to do the analysis and proved that that's not true.

That's awesome. The things that we find that are a little bit surprising. The smart guys at Zillow have put out a book recently talking about the non-intuitive or anti-common sense factors that go into home values. We're validating a lot of that stuff, and a lot of that stuff is questionable. Per dollar of renovation, working on your bathroom tends to do better than your kitchen. Given a neighborhood, square footage tends to be a proxy for a number of beds and baths. There's a bunch of these little anecdotes that come out in the data that are interesting. We don't get to nearest Denny's yet.

What was ZXV?

I had transitioned away from my PhD and was doing more quant work. I had been more distant from the project but I remember you and Tom and I, good mates, getting together to form some sort of corporate entity. Now, at the time, I was thinking I wanted to do more of my work as a contractor. I thought about forming an entity with you and Tom and eventually doing some of my consultancy through that entity. Even though this changed later, I remember at the time you wanted an entity for a specific client project, or was it a money thing?

I honestly forgot.

We were at Victoria Train Station at the pub above the Burger King. And we were having a meeting where we were going through a deck, you had put together because we were going out to Surrey for—was that Getmapping?

Yeah.

We were tempted to do a project with them. This was also around the time we were actually forming the company. We bought an off-the-shelf company called ZXV from one of those brokers. The U.K. quirk is it tends to be easier to do that than found a company, so we bought one and we would go through the re-naming later. You had grand, intelligent ambitions to grow it into a GIS-driven startup. I was a little misaligned, and was thinking of it more as an entity through which to do some consulting. You know, I'm working with my buddies, so if I got professional liability insurance, bank account, naming we could sort of distribute some of that. And at the time, I think Tom was at the architecture firm and still flirting with The Bartlett. That's the beginning of ZXV.

It was like six months later when you got Ben Russell, Nick Black and one other person you wanted to bring in as partners. I didn't want to become a founder of a GIS company. And you knew that, so you asked me to step away, and Tom as well. I remember a meeting that was all tense just because it was one of these unfortunate friends-and-business blurrings, but I remember it was at your place in Camden Town. I think, that's where it transitioned from a Steve-Tom-and-Ben kind of umbrella org to a Steve-Nick Black-going to become CloudMade-soon org.

My memory is that I started the company with you and Tom. And then quit my job but it was difficult to be running a company where you're on your own. And at the same time, although I forgot the timeline exactly, Ben Russell and Mikel Maron and I were trying to figure out if there was something there. And then that didn't really work out either. And then I split the company later with Nick Black.

And Mikel is still an OSM champion. Every couple of months, I seem to hear of something interesting he's doing around GIS in devastated places. I had forgotten he was involved really early.

Yeah, because when Tom and I got away from it, you guys had also, about that time, figured out a way to get some funding and make it a proper startup. That was the Danish VC, right?

Right.

CloudMade's still chugging. I don't know too much of what they're up to lately. From the start, I saw it differently than you and Tom, which is probably a recipe for disaster. It was tense for a while there, which is shitty.

So what should I've asked you that I haven't?

When you're a young fired up grad student, I think the activist side of OSM was very attractive, and it wasn't just for meeting women. It was having a big bad guy, even though in hindsight that was incredibly simplistic. The Ordnance Survey as a proprietary locker-up of the data was a good thing to push against. It gave us a fight.

A lot of the early energy for OSM, in addition to being a wiki of maps which is fucking cool, from a tech perspective -- I think the other side is that activist side. I remember that being attractive early on because I'm a leftie from Chicago. I've marched many times in my life for and against various causes, so the political side of things for me was important -- especially thinking ahead to the metadata and giving a real Ben Russell's HeadMap Manifesto—style source for the world.

It does. I mean, chasing the big bad guy was a very conscious move and it really helped unify everyone, right? Let's all fight. And there must be some deep irony somewhere that the big bad guy was actually funded by our taxes. We're actually paying to fight ourselves, you know?

Go government. The other anecdote I would bring up is What the Hack?

Why don't we talk about that? Go on.

It was one of the first times I spoke about OpenStreetMap to an audience outside of the U.K. You and I went to an underground hacker conference in Holland in '05 or '06? It was pretty early. It was called What the Hack? And it was held in a field in a small Dutch town called Boxtel.

What the Hack? fancied itself a kind of Chaos Computer Club conference, coming from the same roots. I remember you and I were there, camping in the field. I remember very post-ironic rave moments of dancing in a tent with a bunch of other nerds to bad techno. The talk itself was pretty cool because this again hit the activist theme. I mean, we had a big bad guy! We were giving people accurate data where before it had been obscured and tucked away. We gave our typical technical audience spiel on OpenStreetMap in a giant tent in the middle of the sticks in Holland. We ran into a couple of white hat guys working in London. We never stayed in touch, but there were a lot of stories.

Yeah, so my memory of both What the Hack and then also the Dorkbot events is that it was sort of anything was possible. So you'd go to Dorkbot and there'd be someone showing a new poster or something and then someone who had made like a 50,000-kV Tesla coil, right? And then I remember when we went to What the Hack?, I remember one of the tents near us. A guy brought his tent and then a 200-foot power cord and his desktop computer with a CRT monitor and was

sitting next to his tent with a desktop computer with a large keyboard and a monitor, just like you'd be sitting at a desk at work, right? Hacking on codes. I'm sure that there would be more fruitful ways—more comfortable ways of hacking on your latest project than sitting in a field and ignoring everybody and working on this stuff, right? And then there were people just building all kinds of random things and it was really very much this energy, you know?

It was an interesting era. Because it was '05, so we're well before this current tech bubble. We're long after the first NASDAQ bubble. Pre-Twitter being the kind of demonstrable social network that can empower a large group of people. This is long before the Chinese peer-to-peer mesh networks between phones. It was the cusp of big tech developments, and from a personal perspective you and I were still sort of like "grad school." We hadn't gotten too serious about our jobs yet. I don't want to look back on this as an old man, because I still sense that energy when I go to some meetups in the Bay Area right now. It's not gone but it's very different. It was just a rougher kind of tech. There were no standards for a lot of things. Accelerometers on every gadget were not defaults.

One of my grad school projects that never came to fruition was using accelerometers for performance, but that stuff was just too expensive. I think it was a different generation of tech where things were very unstandardized and messy, to a fault. I mean, it was harder to get anything done.

Sean Gorman

Bio

Sean is a founder of Timbr.io - a platform to enable a community repository of re-purposable algorithms and data. Before starting Timbr.io he was the founder and CEO of GeoIQ - a collaborative data and analytics company serving commercial and government customers. GeoIQ was subsequently acquired by ESRI where Sean worked integrating social data with ESRI's mapping technologies. Sean has also previously worked in academia serving as a research professor at George Mason University. His academic research was focused on the intersection of complexity science, statistical mechanics and spatial analysis. Sean received his PhD from George Mason University as the Provost's High Potential Research Candidate, Fisher Prize winner and an INFORMS Dissertation Prize recipient.

Favorite Map?

"The map of roads and land use in Kabul, Afghanistan. We worked on a project to open up Afghan government data for use in OSM, and the map has always been a favorite."

Interview

What I'd like to do is sort of go chronologically. I think that makes the most sense. So I want to start with who you are and what you were doing before we met at CASA.

Basically, I was getting a Master's degree in Geography at the University of Florida in the late '90s—right about the time the dot com and telecom bubbles were kicking in. And during my very first semester, there was a big article in the Wall Street Journal about IBM putting $1 billion into the Internet. Being in geography, I was like, "Hey, is there a geography to the Internet?" It was '97. Shortly thereafter, people started coming out with things saying the internet was going to be the end of geography, the death of distance et cetera. So I got interested in the question, the obvious place to start was looking at the North American Network Operating Group list. They were the people who ran the backbone of the internet. We passed some messages back and forth, and then researched other projects online that were looking at the geography of the Internet.

That's where I came across Martin Dodge. He had started the Atlas of Cyberspace out of the University College London with CASA. So I e-mailed him and said, "Hey, I'm starting to look at where the fiber optic backbones and routers and physical infrastructure that runs the internet, and seeing if I can parse out some kind of geography from that." So he e-mailed back and gave me lots of good recommendations of resources to check out. Then we started e-mailing back and forth after that. I'd share things I found. Martin would share stuff that he found. And then we met another researcher, Matt Zook, who was doing work at Cal Berkeley on mapping out domain names. Then we met another researcher, Anthony Townsend, up at NYU who was trying to establish a free wireless network and mapping out Wi-Fi spots in New York. All this was in '97 to '98 timeframe.

The four of us started working together on various things. Eventually Martin would invite me to come out to CASA to do some work. Martin came over to D.C. to work with TeleGeography for six months to nine months, something like that. By that time, I had left Florida and I was up in D.C. doing a startup. Then eventually back at George Mason to do a PhD. The four of us started collaborating on more and more things. I took a couple of trips out to CASA to hang out and that's actually where I first met you, Steve.

I think we met at one of the pubs. Correct me if I'm wrong as I'm going through this—you were working with Martin because Martin was also acting as the I.T. person for CASA. I think actually before we met at the pub, he had sent an email because you were doing research on fractal dimensions of trace routes. We'd been doing a lot of trace route work on top of the fiber optic data that we were mapping out.

At the time we'd collecting a large amount of fiber optic data by horse-trading on the North American Network Operators' Group (NANOG) listserv. For example, we'd find an AT&T network and then we would trade with the MCI backbone operator because she had a map and was willing to swap. In short, "Hey, we'll give you AT&T for MCI." Then we just started building out more and more of these and then created a GIS database where all these long haul fiber routes were. Then we started doing metro fiber, followed by colocation facilities, carrier hotels and things along those lines, like trans-oceanic cables. We just grabbed everything that we could get our hands on. Remember this was all in the '90s, pre-9/11, so there weren't really any security concerns. People were excited to share that the Internet was everywhere but it really wasn't everywhere. It was just in big cities, at least when it came to big data pipes like the OC 192 fiber cables.

Martin still continued to do various cool things around the geography of cyberspace. CASA was just an awesome place back then. There were lots of really smart and cool people working there—you and Paul Torrens and Naru Shiode and Tom Carden. Yeah, I was super happy just to get to interact with such a brilliant group.

After 9/11, George Mason got an earmark from Congress to set up a critical infrastructure research center out of the law school of all places. So we submitted a proposal saying, "Hey, we have all this data we've aggregated." Our original work was trying to show that fiber optic build outs led to economic development—that basically the more fiber you had connected into your city, the more economic growth that it could create. That was our hypothesis. So we sent this proposal saying that we have a few ideas of how we would look at fiber infrastructure aggregations from a vulnerability standpoint, but this was not at all what we built it for. Then new GMU center asked us to come in and give them a briefing on it.

They asked, "You actually have this data? It's not dummy made-up data, trying to test an algorithm?" We said, "No, it's the actual data. When the folks at the research center saw it, they took us to go brief Richard Clarke who was kind of a cyber czar out of the White House back then. Then he sent us to the NSA and all the other three-letter agencies to show them what we had. They kind of freaked out and said, "If this was our data, it would be classified and nobody'd be able to see it but since you created it from open source, there's really not anything we can do because it's all public domain data that you've just aggregated in interesting ways." Word got around that all these agencies were kind of freaked out by the briefings we were giving. A Washington Post reporter heard about it, came and spent like a week hanging out with us and seeing what we did and our research. And then we didn't hear anything for like a month.

And so, I headed out to CASA that July to do work. I think that's the time that we actually met up in the pub. At the end of that trip, all of a sudden, the Washington Post published the story and they put it on the front page of the paper. It was a super slow news day. And then CNN wanted to interview us. And so a black Mercedes from CNN came and picked Martin and me up and took us over to the studio. We did an interview with Wolf Blitzer which was kind of crazy at the time. And then that started off slow news day spirals and more news stations get interested because there's nothing else going on. And so, I flew back to the U.S. the next day and we did a little media tour with all the standard hyperbolic U.S. media places and a few international ones.

Tell me if I'm wrong but my memory is that there were some headlines that they put together that basically if you cut two bridges across the Mississippi, you'd divide the U.S. in two. Is it like that or am I misremembering?

Yeah. That's right. It was more in the Rocky Mountains. There are only so many ways to get across the Rocky Mountains. There are actually much fewer ways to get through the mountains than there are across the Mississippi River. But yeah, it was sensationalist things like that.

So if you blew up I-70 or something?

Yeah. That would take out a lot for sure. There are these different choke points. Like the Hudson Tunnel going into New York is a huge choke point. These are usually caused by geography—there are only so many ways to get across that particular piece of geography, and then economies of scale aggregates everybody onto that least cost path. Then you get a bottleneck of vulnerability going through least cost path along a single route.

Right, and so you can sort of imagine there's a lot of different cables and fibers going through these particular choke points, right?

They pretty much all follow roads for the most part. A few things follow gas pipelines and electrical power conduits but they're pretty much all street right of ways.

Not railways?

Yeah, railways also but the vast majority, especially metropolitan fiber. The internet is buried under our streets or strung along poles. There is a big ditch next to the road or they dig it directly under the road. Such as when you see a lot of the spray paints on the street with arrows and that is where the fiber is. They usually would cut it or they'll drop it into the manholes.

Okay, so when was your first brush with OpenStreetMap?

I think we talked about it at CASA when I was there that summer. Maybe the idea was kicking around then. Although I'm vague on the details other than talking about it at the pub or my misremembering because that was summer of 2003 but I can't remember where that was, in the arc of the project.

The domains were registered in August 2004, so it's about a year prior?

Yeah maybe we were mostly talking about fractals, bit torrent downloads, and I thought there was a little bit of the idea of OSM but I may have falsely imposed that into my brain associating it.

What were your initial thoughts?

I thought it was really cool. The funniest thing I think about my initial reaction to OSM was that when I saw OSM going up, we had spun a company out of George Mason, GeoCommons, and I remember seeing OSM things pop up in my newsfeed. But I totally did not put two and two together that Steve Coast and OSM were associated—that ideas we had talked about as just early concepts of crowdsourcing or getting infrastructure data and how to collect data and those conversations—that any of that had any connection at all to OSM.

It wasn't actually until one of the first Where 2.0s that I saw "Steve Coast" was talking about OSM. I was like, "Oh wait, is that the same Steve Coast that was working with Martin at CASA?" Then I went and looked it up. I went, "Shit it is. It's the same person." So it was really interesting from that standpoint. I'd seen that the project looked very cool and people were starting to talk about it, but it wasn't until Where 2.0 that I realized that it was something that had partially sprung out of group in CASA I loved hanging out with. That was a really cool "Aha!" moment for me. The funny thing is, we had started doing crowd sourcing with GeoCommons but a totally different kind of crowdsourcing from OpenStreetMap's.

Do you want to talk about what you were trying to do with GeoCommons?

Sure, as we went through and briefed a lot of these different government agencies, everybody began to ask us about the provenance of the data. We'd informally collected our data from a social network of data holders that we incentivized with our pool of data we shared back. This resulted in people always having questions of, "Was it authoritative data?" "How do we put GIS metadata around it?" Although everybody agreed it was the best data that was out there –mostly directly from the network providers themselves. Experts that knew a lot about telecom went in and looked at it, and everybody agreed that it was really high-quality data. Problem was it didn't go through any of the traditional processes of putting that data together. We also had a whole faction of stakeholders who wanted to bottle all that data up and make it go away. They thought it was more of a risk being out in the public. Instead we should just burn it, get rid of it.

So the typical security by obscurity?

Yeah, stick your head in the sand, ostrich-style. We of course had the opposite reaction, saying, "Well, no. Now that we know that this is out there we can begin to address the vulnerabilities and figure out how to address them." If a grad student can do this in their spare time, then you can pretty much assume that the bad guys will be able to figure it out also. Security through resiliency is way better than through obscurity. Basically, our concept and response was to come up with a place where systematically people could contribute data and re-mix it and re-purpose it the same way we had done manually. The idea was to create a social network of data contributors.

We ran into the same data sharing challenges when we started doing emergency response after Hurricane Katrina. We also did work trying to help out with the London bombings. I'd call up Martin and say, "Hey, do you have census data for London?" and he goes, "Yeah, I have some files sitting on my computer with census data for London." then he would send it over and we would call up somebody else to see if we could get LandScan data. They'd have to courier over DVD's. Then we had to amalgamate all the data together but it took a huge amount of time. The same thing with Hurricane Katrina, trying to get pipeline data, wind surge and storm surge data.

As a result we thought it would be a lot easier if there were already a place where people contribute their data sets, and then when a disaster happens or humanitarian crisis people can go and find the data that they need or contribute data as it gets created on the ground. That was fundamentally how the concept of GeoCommons came about. If we could have a common place to contribute data and then folks could re-mix and re-mash that data to create maps and analyses.

So is it fair to say GIS on the web?

Yeah, GIS with a social dimension – a crowdsourcing dimension to it. For instance, OpenStreetMap is focused on how you create a single data set with a lot of people contributing individual parts – the roads, the intersections and the points of interests to create one big data set. Whereas GeoCommons was more about how you re-mix a whole bunch of disparate types of data sets to put on top of an OSM base map. Alternatively, can we remix OSM data—pulling data from OpenStreetMap and adding Census data or EPA data to create hybrid data sets?

The interesting aspect to classifying GeoCommons as "GIS" now is that when we launched it, a lot of GIS people had a really negative reaction. I remember there a blog debate between James Fee and Steven Citron-Pousty on whether it was crap or useful. James Fee stated, "Oh, there's no good quality metadata in there. I would never use it." Then Steve was like, "Oh, there's heat maps and there's all sorts of interesting data in there. I think it would be useful". So it was really interesting because a lot of the traditional GIS folks weren't big fans of it initially or the concept of crowdsourcing. Maybe they still aren't.

Do you think they were worried about their jobs?

I don't know. I think it's kind of this general GIS reaction to things. Everything needs to be very structured, professional, and official, with a lot of pedigree in order for it to be GIS. It is a continuation of that whole professionalization, accreditation culture meme that surrounds GIS. The meme has served GIS well historically, but I think when a lot of this early crowdsourced data came out, they had a very allergic reaction. It didn't come from the GIS community and it didn't have the GIS trappings to it.

Early on, we tried to start a conversation with GIS folks saying, "Well, what metadata do you need?" Because 452 elements of FGDC metadata or however many there are—is ridiculous. Nobody's ever going to attach that to their data when you're crowdsourcing, and getting people to donate data. We ended up using Dublin core metadata, which is, I don't know, 12 or 14 elements and then mapped that to FGDC and ISO. That seemed to make people pretty happy, but I think, there's multiple vectors. OpenStreetMap was growing really strongly and other folks were pushing stuff out there that didn't have all the GIS overhead to it. People were becoming more receptive. It's been fascinating to see how that's happened over time. To the point that now it's almost that there is de facto acceptance of crowdsourcing.

Sean Gorman
So it's like GeoCommons was unlocking existing data that was spread in all kinds of different places? And OpenStreetMap was creating new data by getting completely, for the most part, amateur volunteers?

Mm-hmm.

So, they're different on more than one level, right?

Yeah, definitely. That's a really good kind of encapsulation and differentiation. I mean, OSM was a much bigger cognitive leap for folks than GeoCommons was. Since GeoCommons was just, in many ways, creating a more efficient way to pool the data they were already working with. It also happened to unlock a lot of data sets that GIS people had never thought of using or working with. For example bringing in GeoIP address logs, social data and all the interesting things people were beginning to scrape, grab and build, but had never been in the GIS world. That said, the substantial base of it was government open data sources that were just sitting around on people's hard drives, then trying to aggregate those in an interesting way. We never got the same kind of almost viscerally angry reaction that you would get from bringing up OSM at a mapping conference. The NAVTEQ person would get up and rant about accuracy and how nobody would ever use this for mapping or navigation. GeoCommons never had that level of reaction that OSM did.

Yeah, luckily, it didn't work out that way, huh?

Yeah. It's really funny looking back at it now. I can't remember which conferences they were—maybe the Location Intelligence Conference and there were a couple of others where the NAVTEQ, TomTom, and a couple of other ones would get on these really angry rants whenever the topic of OSM came up.

Yeah, it's like first they fight you - that kind of thing. So what was your impression of OpenStreetMap around the time of, I guess, Where 2.0? And then from there, going forward, what was your interaction with OpenStreetMap with GeoCommons?

Mm-hmm. Yeah, we were really excited about OSM because it created a lot of interesting opportunities for us. I think initially, the way we thought about it was tiles, right? You have this open free base map and you don't have the overhead or contractual concerns with Google and Microsoft. With GeoCommons we wanted to give people the opportunity to have as many different base maps as possible. Inevitably we'd get into some interesting kind of legal snafu trying to support the different base map providers.

There was the Mapstraction work on to help you not get locked into tiles from one provider and be able to dynamically swap out base maps. It was definitely still early days. It wasn't clear what you were allowed to do, what you weren't allowed to do, what was derived data when you're putting data on top of the maps and using them for different purposes. As a result our initial interest was finding a default base map we could use that would never get us in trouble. The solution was OSM.

One of the things we did early on was we started working with Tom Carden and Mike Migurski at Stamen to come up with a custom-style on top of OpenStreetMap that was good for thematic data. We came up with this grey base map with the idea of being able to layer into the base map – put the streets and the place names on top of the vectors, a cartographic sandwich effect. Then open sourcing the project so others could create their own custom base maps. The layering was Andrew Turner's idea for a thing he called acetate. We worked a lot with Tom and Mike on coming up with what we collectively thought was a nice-looking style for thematic cartography.

Did you find it funny that that is still effectively the cutting edge that people are still producing those tile sets and selling them?

The wild thing is just how long ago this was. It was 2008 or 2007 maybe the team was doing this work. It really was a great team with partners like Stamen. It was an awesome time with lots of creative people working on geo that were new to the field with exciting ideas. In some areas, we've really evolved and there's been a lot of really cool innovation. In other areas we still seem stuck in a rut. After the tiles we also started working a lot with the data itself. Doing extracts from Open Street Map and just trying to get little portable snapshots that people could download as shapefiles and KML to take into different places. That work really started to ramp up. Initially, we were just working around humanitarian crises. Where if something happened in a place and there wasn't any good street data for it we'd go to OSM and take a capture for that geography. Then put it out in a vector format that people could then download and re-mix or re-match with other stuff. We were doing this in a couple of places but where it really, really took off was in Haiti after the earthquake. Kate Chapman took the lead in reaching out and working with OSM. Trying to connect the dots on how we could get the OSM data distributed to as many people as possible. The road data was really bad in Haiti and OSM does such an awesome job of using the community to build out much more detailed road data really quickly. Kate did an awesome job connecting all that stuff up. Then eventually, she had the opportunity to go do that full time with Humanitarian OpenStreetMap. A big part of work with OSM was through this humanitarian perspective.

Since OpenStreetMap and GeoCommons were being used during humanitarian crises and disasters we collaborated quite a bit. GeoCommons was playing this role of a neutral data repository to stick data and make it available to governments, NGOs, and citizens without anybody feeling weird about where the data was coming from and who was doing the handoffs.

Sean Gorman

Can you talk a little bit about the difference between Katrina and Haiti? And all these tools and how it worked out?

Yeah, definitely. You know, Katrina was interesting because it happened in a place that was very data-rich. There were already a lot of existing data sets around New Orleans, and the panhandle of Florida areas that were hit by the hurricane. It was really a matter of trying to get the data that existed into the hands of the people who needed to do response. The problem was that system was really broken. It was a lot of sneakernetting with DVD's being pushed around and overnighted—just a lot of inefficiencies. People re-doing the same tasks over and over again.

By the time the Haiti earthquake hit we had decent data sharing, but then all of a sudden we went into a place that didn't have data. Responders were really at a loss as to how to handle that. Before Haiti, everybody looked at crowdsourcing, especially in the government and NGO community, very dubiously. Often saying "Okay, if somebody's pushed data into GeoCommons, I know who that person is and I know where they got it from, and so link back to the source and the metadata, I'm cool to use it." On the other hand sites like OSM that are coming purely from the crowd they were really skeptical about using it.

Then over the course of events in Haiti - where there was literally almost no authoritative data, people were forced to use these new data sources. Going to OpenStreetMap, going to Ushahidi and going to these other texting services to get data.

When people saw there wasn't anything else to use, they were forced to use OSM. All of a sudden, they realized it was better than the official source data that they had used in the past. After about two or three weeks, the official sources started to kick in, and data started to come in from the government. Responders started getting this "authoritative" data on the hospitals and the roads and responded, "There's no way we're using this. This data is so much worse than what we're getting out of OpenStreetMap and the crowd sourcing initiatives. Not only is it late. It's already old and it's already inaccurate."

What were people using that data for?

A lot of it was for trying to route search-and-rescue (SAR) folks to where people needed help. Haitians would be texting in that they were trapped in the rubble and then the SAR folks would need to figure out how to get from point A to point B. There weren't any street maps to allow that to happen. That was kind of the first use case that people were trying to solve. The follow-on use cases were then trying to deal with the refugee camps and the hospitals. Where are the refugee camps popping up? Where are the people that need help? Where are the hospitals? So that you can get the people connected to the hospitals.

Then people started to build things--hospitals, medical tents, and related facilities. It was a very dynamic and changing landscape keeping track of all the things that were changing and getting created. What roads were collapsed? What buildings were collapsed? What hospitals were accessible? Which ones were damaged and couldn't be used? What new ones had popped up? Where they had popped up? Where are the refugees in association with these things? All this was super fluid and OpenStreetMap and the other crowd sourcing approaches that people were testing out did a really, really good job of sorting that out. Making the situation a whole lot better than it could've been.

So for most people, they know where their hospital is because it hasn't moved in years, right? But you're in a place where the location of the hospital and whether it's even open is changing on like a daily or hourly basis?

Yeah and you also had an influx of a huge number of non-residents. The number of NGOs that deployed into Haiti, the number of U.S. government support personnel that came, United Nations folks that came was enormous. None of these people knew where anything was and there weren't any street maps. Plus, the landscape was fluid and changing. They needed a baseline and they also needed change sets on top of that baseline. OpenStreetMap was explicitly built to do that, so it was just a really great fit. During the disaster and especially on the backside of it, all of a sudden, there were no longer any questions about crowdsourcing from a standpoint of is this something we should invest in? Is it something that we should try? That we should use? That we should take off the shelf? Everybody involved was on board and promoting those kinds of things just became worlds easier.

You talked about the sort of the humanitarian angle. Do you think there had been any similar shifts in academia and its view on crowd sourcing and OSM in particular?

Yeah, it's a good question. After the Haiti earthquake I ended up working with Matt Zook at the University of Kentucky and Mark Graham of the Oxford Internet Institute. We wrote a paper on crowdsourcing during the Haiti crisis and all the different crowdsourcing tools that were used and what kind of effect they had. That was just one paper of many that was beginning to talk about crowdsourcing and its effectiveness. People started to build more theoretical structures around how OSM and crowdsourcing fits into geography and social science as a discipline. Even academics were talking about crowdsourcing from a computer science perspective.

It also helped that Haiti and several other events put crowdsourcing and OSM in the news quite a bit. People started hearing about it through mainstream media. I'm trying to think of—I know there were other events where Open Street Map got really good press but at least from myself being directly involved, this was the one where it really became more of a household name for folks. Then people began to realize how useful it was for day-to-day things. It wasn't necessarily just there to just be used in a crisis. There are all these other wonderful ways to use OSM for a wide variety of scenarios and the same benefits that you'd find in a crisis are the same benefits that accrue on a daily basis as far as --that the bridge was closed in Texas after a hurricane and then Yahoo or somebody had finally added it as being closed right around the time they had reconstructed it. So you started to get more of these anecdotes popping up in the media of how major map providers were out of date whereas OSM was almost always dynamically up-to-date.

So it's sort of come for the crisis, stay for the data?

Yeah that's a much pithier way to say it than my ramblings.

No, it's all good. So it feels like it has this progression. Sort of commercially, we talked about academia and then in humanitarian circles about the worth of using things like OpenStreetMap and crowdsourcing, do you think there were also interactions between the three?

Yeah, definitely. I mean, from my own personal perspective just because we'd come out of academia and we were doing a lot of disaster and humanitarian work. Then we also had a lot of commercial customers that we ended up doing more and more OSM-related work with all three of those entities and sometimes they intersected like the academic folks being interested in doing research around how OSM was used in disasters. Then commercial folks being interested in seeing a bunch of news about OSM during the disaster and saying, "Hey, is there a way we can leverage this for doing commercial things?" We definitely saw a lot of those kind of positive externalities happening in that timeframe.

Right and was there a moment, you think, when you realized that OpenStreetMap was going to be good enough? Presumably it was prior to all of this?

Yeah, I think a lot of it was going to those early OSM State of the Map Conferences, the very first one in Manchester that Martin kind of arranged to host. After going to that and seeing all the stuff that the community was doing I was really excited. I think, even as early as Manchester, for me, it's saying, "Hey, this is not only a cool project conceptually, there is a lot of really invested individuals that are doing very cool things with it." In addition they're not all the same thing. There were a lot of different perspectives and ways that people were trying to solve problems. Then seeing how that grew from going from Manchester to—was it Limerick that was next? It all gets a bit fuzzy for me but it seems like every time I went, it got a lot bigger, right? The first time it was, there may have been 100 people at Manchester.

I think it was 80.

Yeah and then all of a sudden, there's more folks and everybody's doing interesting things with it and just the variety and diversity of the group was growing really rapidly. For me, there were still a lot of problems to solve at that point, but you could see the community and how excited they were and invested in it. You could tell they were going to figure out whatever comes down the pipe. So for me, that was kind of the Aha moment when I thought this was really going to be something significant. Then those opportunities arose on the backside of Haiti and other situations, and then OSM really got a lot of attention. That seemed to be where the hockey stick kicked up as far as the number of users and contributors and the quality of the data and the number of different uses. It suddenly became really exciting.

What effect do you think Open Street Map had on other old and new mapping organizations? And I'm thinking everything from NAVTEQ, TeleAtlas, to Google?

It is interesting I think there were three different reactions. One was a hostile reaction of "This is totally different in the way we do things," and it was almost antagonistic. This was true especially early on. Then there were a lot of folks that looked at it and said, "Hey, this is an opportunity for us. We can work with the community and integrate it into what we're doing and I think there'll be a lot of opportunities." Then there was kind of a third camp of "we can emulate or copy and use this concept in a similar different way." What was really fascinating was to watch the first antagonistic camp all suddenly realize that this wasn't something they could ignore and just throw a random disparaging comments at. They were actually going to have to either join one of those other two camps. Come up with an alternative siloed off to themselves, or work with the community and try to figure out how to use that to their advantage as a business.

It's funny, I remember after giving a talk once, someone from a major mapping company came up to me afterwards and said—the first words out of his mouth were, "We can't work with you." Like, "Hi, I'm Steve." That guy introduced himself and was explaining that because of the license, they couldn't import our data because it would infect their data and they didn't want to give us any data, right? And I said, "Well, we're not forcing you to work with us. It's okay." And then I said, "You know, one of the ways you can get involved is that you've probably got backups of your map data from like a year ago or five years ago, right?" and I said, "So what is the value of the data from five years ago?" And the guy said, "There's no value to it at all, right?" So I said, "Okay, well, why don't you donate it to the project?" And he's like, "Oh, actually, it's incredibly valuable". So it's been interesting to see their reactions.

Also, it's interesting to see how their reactions changed over time. I think how the early feelings about OSM from the traditional vendors evolved to today is pretty amazing.

Sean Gorman

What has been the most surprising thing about OSM for you?

I think it was how quickly it grew once it hit that inflection point. You know what I mean? Sitting in those—the early State of the Map Conferences, seeing what people were doing and whether there's 80, 150, 200, 250 folks. Then OSM went from this small group of folks where you could recognize most everyone—to all of a sudden being this massive community of contributors. Now the majority of the OSM community you will probably never see in your life and have no idea to what drives them or why they're participating. I think what really surprised me was when it hit that exponential curve and the growth of the community from this small intimate thing to this huge and massively successful thing. The cool thing is there is still a State of the Map and all the same faces are still there, right? There are always new faces dropping in. Sometimes they fall out, sometimes they stick around but it seems like that that core community is still there pushing things forward.

If you could improve something about the project, what would it be?

I don't know. The thing that I would say that I'd most like to see improved is probably the hardest thing to improve. How do you get rid of kind of the vitriol and flame wars in the community? It's interesting, because in the early days, there were so much antagonism coming externally from the traditional providers and everybody was on a cohesive front that OSM is doing something totally new and different. Now that it's become successful, it's like all that antagonism and vitriol is now inside of the community, arguing over old ways versus new ways, this side of things and that side of things. Sometimes, it seems like there is arguing for the sake of arguing. I think if there's one to thing to change, if we could get back the original sense of community from those early days—really focusing on solving problems and changing the established way of doing things. Then again this may be a false positive generated by looking at the OSM listserv as too much of a key indicator.

Right. So let's jump back to GeoIQ. Let's talk about like how it went from GeoCommons to GeoIQ to ESRI. And if along the way, if there were any fundamental changes like business model, location, who was involved?

Yeah, it was an interesting process. When we originally spun out of George Mason University, our idea was let's do vulnerability analyses for folks based on all this data that we had. So we had some bank customers. I think we did some work for McDonald's. What's your exposure to a disaster based on your fiber diversity and where are you getting your infrastructure from? How resilient are you as an organization? The problem with that ended up being—unless there was a disaster or a terrorist attack, people really weren't interested. They only care when there's a problem and nobody wants to invest ahead of time for a problem that may not ever occur.

This is all in the post-9/11 environment. One thing is, we collectively as a group didn't want to be in the fear mongering business. We felt, in order to sell our product, we had to convince people that everything was terrible. The more disasters and terrorist attacks that occurred, the better the business did which just seemed like not a good business to be in. It's not something that we enjoy.

But what people ended up being really interested in is how we used the Google Maps API really early on to add heat maps on top of imagery. This was right after the Google Maps API had come out in 2005. So we started to say, "Hey, if you have a bunch of points or vectors, we can create a heat map with it." People were super excited about that. We had all sorts of people asking us, "Oh, how can I add analysis on top of Google Maps or Google Earth?" Internally we thought, "Hey, well, maybe we can do something around that." That's when the idea with GeoCommons came along with the thought "Well, you can't do analysis until you have data, so we need data."

This is also right around the time Hurricane Katrina hit in 2005. So, we started to see problems around trying to share data, and how do you get a good pool of data for doing analysis. GeoCommons emerged out that crucible and we ditched the vulnerability analysis work. We just focused on building a Geo platform. At the time, the company was called FortiusOne because Fortius is Latin for stronger. We wanted to help create societal resiliency, make the customers stronger kind of thing. Then we built GeoCommons and eventually changed the companies name to GeoIQ.

It's one of these bad Marketing 101 things where we had also come up with the heat mapping API and we've called that GeoIQ. So we had three names running around – GeoCommons, GeoIQ, and Fortius One. Nobody knew what to call us. So eventually, we got rid of FortiusOne, made the name of the company GeoIQ and then our open-data project was GeoCommons. From that point, we decided to focus on creating a Geo platform, and that was consistent through the history of the company going forward.

We did a first version that we built on top of MySQL. We outstripped the ability of MySQL to handle the amount of heterogeneous data users were pushing into it. So, we switched to PostGIS. Then we outstripped PostGIS and it couldn't handle it. This was 2006 or so. Relational databases weren't hacking it so we ended up building a key value data object store. The current NoSQL technologies weren't around back then so Pramukta built this really simple object store of serialized data. It put data in a queue and pulled it up into memory when we needed it and pushed the older data to disk.

We had a previous version of GeoCommons that we pulled down and then we came out with this new one, built on the object store. That is also when we added more official metadata. Then we added an extract, transform, load (ETL) engine to get data out in a variety of different formats. We ran that for a little while, then we added the map making abilities on top of it.

So the first bit that managed all the data was Finder. Then we created Maker to generate the maps. That's when we started using Open Street Map as our default base map. It is also when we worked with Stamen to come up with a custom version of OpenStreetMap specifically for doing thematic mapping. Then it gets kind of boring from there, but that was already boring to be honest.

How did it go from there? It's up to you what level you want to talk about this but I think VC shenanigans happen a lot more than people think and it's always suppressed because it's never in anyone's interest to talk about it, right? So it's up to you. If you can, I'd like you to talk about misaligned incentives. It might also be better to sort of back up and talk a bit about how the company was funded in the first place because we haven't covered that.

Yeah, definitely. When we spun out of George Mason, I had a friend of mine that worked at the University of Maryland helping spin companies out of the university. I asked him if he could help. The patent lawyers had shown up to our lab at GMU. I didn't know what to do and he offered to get breakfast and provide some advice. It ended up that the University of Maryland also had a venture fund that he worked with. They became interested in what we were doing and they put in a seed investment.

Their fund was this Small Business Administration and the University of Maryland Public-Private venture. Then we shifted gears into doing the Geo platform stuff and raised an A round. That was led by In-Q-Tel, which is a venture fund for the government agencies. They had just sold Keyhole to Google the year before. So, they were really interested in what's going to be next for Geo. What do you do after you zoom in on your house on Google Earth? What are practical things that you could do with these kinds of platforms? We had a technology for getting data on top of Google Maps as well as doing analytics.

In-Q-Tel got interested. They led the round. There's also a New York investor and some local D.C. folks that came in on it also. Everything was good.

We went through like four or five CEOs in the history of the company. So we'd had a CEO from the seed round through the A round. That person didn't work out for a variety of reasons that are all under non-disclosure. This happened when we were in the middle of fund raising. We had basically run out of money and so I was going around town, pitching the company. In-Q-Tel got interested. Then other folks got interested but we didn't have a CEO. Eventually the partner at In-Q-Tel said, "Well, why don't you be the CEO so we can get the deal done?" Under the rationale "we have the deal ready to go but we still need to recruit a CEO once we get the deal done." Basically, the folks that were investing said, "But we don't know who that person is. We know who Sean is and we like Sean so why don't we just make him the CEO?" So that was how I ended up becoming the CEO of the company but with the idea we would recruit somebody that had more experience that wasn't just a grad student trying to figure stuff out.

I think you do yourself a disservice to yourself.

Well, I would also say, for those who are ever interested in doing startups, that a lot of inexperience can be trumped by the fact that you're more invested in the success of the start up than anybody you're going to hire to run it. You can figure out a lot of stuff on your own or with help. Get good advisors and don't underestimate what you can accomplish with your founders. A lot of folks coming in later or through VC's have their own agendas, and they can't figure everything out just reading TechCrunch.

Anyway, we had a couple of other CEOs that came through. Some were better than others but eventually, at the end of the day, we got to the point where some of the investors were looking for a big home run because of what was happening in their funds. The last CEO was brought in to figure out how to create this home run so they could get a big exit. Basically, they wanted to recap the company. I won't get into the technical bits of it but they wanted to take the company in a new direction. That changed from time to time but the main thrust of it is they wanted to go into geo-couponing because GroupOn and Living Social were in TechCrunch and getting a lot of buzz.

Chasing tail lights.

Yeah, we called it management by TechCrunch but they're basically wanting to recapitalize, to use the technology base and go on to one of these exciting markets where a lot of investment was flowing. Then try to raise a really big round and have a huge exit. Their premise was that our current business plan around GeoCommons and what we were doing with it had failed and it wasn't ever going to go anywhere. In the middle of this debate, a very large Geo provider came in and said they were interested in acquiring the company. Their corporate development folks got in touch with me and said they were interested in doing an acquisition. This was kind of the holy grail of acquisition outcomes for us as far as the company, and a lot of what we built the company conceptually for doing. The reason In-Q-Tel had invested.

Long story, short, the big investor felt like they could get an even bigger home run with the pivot into geo-couponing or whatever it was going to end up being. For a variety of reasons, on both sides, the deal ended up falling through. So the acquisition ended up not happening but at that point, we had two camps of investors. One camp that wanted to do an exit now based on the company that had already been built. A second camp that wanted to try to do this recapitalization and go into a new direction with a new team. So, that made for a fairly stressful situation and a lot of typical soap opera startup things that happened to many founders so it's totally not unique.

We were working really hard on trying to get the current business sold. There were a couple of close calls with different companies that were interested in doing the acquisition. Then it ended up at the 11th hour being ESRI. We had another offer that we thought was going to happen and then ESRI came in and countered at the last minute. We ended up getting bought by ESRI which was a great outcome for the team, especially since the alternative was getting recapped and most of us going away and it becoming a geo-couponing or whatever company.

It was a challenging situation from several perspectives. In many ways, ESRI was an exciting opportunity because there were all of these things that we had been trying to innovate on in GIS, and here were the folks that had invented GIS. We had the opportunity to collaborate with them—to come up with potentially what's the next evolution or the next generation of GIS. It was the chance to be able to help out with that little bit. It was also awkward, in that ESRI had been our primary competitor for the past six years. We were excited but also a little worried at how that would work out. ESRI ended up being very accommodating and open to the team coming on board. I think with a lot of our stereotypes about ESRI we were mistaken, and ended up finding a lot of people that were really excited about geography and making maps. That all worked out really well.

I ended up not staying after my first year for a variety of reasons, probably the biggest being that my wife had gotten a job at the University of Virginia and I was commuting from Charlottesville up to D.C. For me, personally, I fit a lot better into a startup than a big corporation. So we went our separate ways and they've continued to do awesome stuff with the team that's still there. Various folks that have left have gone on to do really cool things also. As far as it comes to GeoIQ, I think the thing at least that I'm proudest of is how successful the alumni from GeoIQ had been, doing a wide variety of different things. And all of them have gone on to bigger and better stuff so I think that's probably one of the best outcomes that can come out of a start up venture.

It's interesting when you're trying to sell a company that your investors can have slightly different alignment because they're looking for an almost unrealistic amount of money out of it whereas some lower figure would make most of the other people on the cap table happy but they put all the money in so they have a large share of the influence.

Yeah. That was a huge learning opportunity for me. It was that a lot of times these things don't come down to what you personally would consider rational accounting because your company's performance is only one variable in an entire portfolio of companies that they've invested in. In some venture funds, a base hit where the company sells for $30, $40, $20 million is great. Other companies, for a portfolio that's not done as well, it becomes a drop in the bucket and they really need a home run in order for the portfolio to do well overall. And i`ts also how much dry powder, as the VC say or how much cash is in that fund? How well can they continue to fund you through the lifecycle of the company?

So we had some folks that, you know, would run out of dry powder so to speak or they'd have limitations on how much money they could put into a company which then allows the other company to subsume that chunk of the investment which is called going super pro-rata, which then they get more control of the company than maybe you thought would have been impossible when you began, which can give a company controlling interest and give them a lot more leverage in deals and the direction of the company overall. So yeah, there's just a ton of nuances like that which I had no idea about before going into this, that we learned going through the process of it which, you know, unfortunately a lot of these things you can't learn except for by doing. But hopefully, in the future, as more and more of these kind of things come out, people get a better idea ahead of time. But sometimes, it's unavoidable.

Because we sort of went into some of the technology of GeoCommons, do you have any thoughts on the technology behind OpenStreetMap, like the tags, the API and all those things?

Yeah, it's interesting. Actually, I got a question on this very recently from a group, a large enterprise, which was really interested in implementing an OSM-type system for data collection and crowdsourcing. They were trying to decide between the OSM stack and some of the stuff that's being done from "a big data perspective," like Acumulo. It's one of these interesting things where it's amazing to me how well a very pragmatic solution with the existing components can scale and be successful for a very long time. Even though the current approach works incredibly well, people still want to go with the latest thing even though it may not be needed at all, just because.

That was one of the fascinating things for me with OSM was how a really simple set of principles, on what was a sophisticated technology stack at the time has evolved and scaled into something that has gotten to the size that it has. I don't know nearly enough currently on where things are sitting and how far that infrastructure will be able to go, but that side of it has always fascinated me. Although I have not spent a lot of time under the hood, by any stretch of the imagination, looking at it. I just love the idea that a simple set of concepts being able to scale because they're simple - versus trying to put all the complexity up front. I've seen lots of Geo projects come up with sophisticated ontologies and schemas for how you're going to manage all these different data types and values. They end up with this huge, complex, crazy beast but it never is able to grow and ultimately fails. Where OSM did the opposite of that and is going incredibly well.

What should I have asked you that I haven't asked you?

Sean Gorman

I don't know. Why wasn't I more involved in OpenStreetMap?

Why weren't you more involved in Open Street Map?

Good question! When you asked about doing the interview about the book – I definitely thought "why me?" I've always felt like somebody that was sitting on the sidelines. I may very well be the least involved person on the list for this book. A big reason I was less involved was we were putting a lot of time into a complimentary project and that ate up a lot of resources. In hindsight, I wish I had gotten more involved than I did especially on kind of the nuts and bolts of participating with creating things versus managing things.

I asked the question but I really don't know why I didn't invest more time than I did. I think a lot of it is time being a precious commodity and my personal skills probably not being super helpful, compared to the folks that were working at GeoIQ—whose skills could've been a lot more helpful. So giving Kate time to work on stuff, giving Andrew time to work on stuff. Giving the GeoIQ team time within our task or giving them direct OSM tasks, or funding things they could open source that would help out OpenStreetMap. Those behind the scenes efforts seemed to be more valuable, or maybe that's just my excuse.

So what I thought you were going to say was "Because I was working on the secret government version of OpenStreetMap, Steve." Are you allowed to talk about any of that?

Actually I didn't have a security clearance the whole time I was working at GeoIQ. I'd always avoided it because we wanted to be able to talk about anything open source data without feeling there was any kind of repercussions that we would get us in trouble. Everybody had always told me if you had a clearance your work would get classified and I'd never be able to talk about it. So it's like, "Well, I never want to get a clearance because then I'll never be able to talk about anything." I can pretty much talk about anything during that time period for sure. But was there something specific?

No, but I've heard stories in the past from people that there's an OSM stack internally used by security services, and that it has more data and it's wonderful and stuff. But I mean that's literally the extent of my knowledge.

Yeah and I don't think there's anything secret as far as the government communities using Open Street Map data. I know some of the best data in a lot of places that weren't initially covered in detail by the commercial providers came from OSM. We did a lot of work with a friend, Todd Huffman, who was doing humanitarian work on the ground in Jalalabad and worked with local youths to map the area. Obviously, the government folks were really interested in that data and people definitively used the data. I don't know if there is work going on where people are adding to those databases without contributing back. I never actually go in and look at any of that stuff.

I do know that there are people in that community that would very much like to contribute back to OSM because they're very appreciative of all the good quality data that they get. I have been part of discussions where groups have discussed about how the government community—especially that nexus between humanitarian work, the military community and the NGOs—can contribute data back.

Actually, I've heard information about that in the last week or so. I do know that the problems are the bureaucracy and regulation government users have to try to figure out. There are people in that community that are actively trying to figure out how can they contribute back to OSM with some of the data that they create internally because they don't have the issues like the big map provider who says, "We can't work with you because we can't contaminate our data." Public domain data once it becomes public domain is great.

So hopefully we'll see some of that in the future. Generally, my conclusion from having worked in that space tangentially is that it's not nearly as Hollywood as everybody would like to think. It's usually far more mundane and boring and not exciting at all. It's also never nearly as nefarious as people think. Obviously, there have been abuses. The NSA issues that have been in the news are not good by any stretch of imagination. It's always hard to say because when you work with individuals on a day-to-day basis, they're individual personalities, right? They're all out there just trying to do the best thing they can to solve the problems that they have in front of them. And there is there's not a Mr. Burns sitting there tapping his or her fingers together with a master plan of how we're going to hose everybody. Sometimes, somewhere along the line, somebody loses the ball and things go sideways and there needs to be transparency and accountability for it. Obviously, there's a lot of work to be done there but I don't know I'm rambling again.

Muki Haklay

Bio

Muki Haklay is a Professor of Geographic Information Science at University College London (UCL). He is researching issues of public access, use and creation of environmental information and therefore interested in participatory mapping and citizen science. He learned about OpenStreetMap early on, and, with the support of the Royal Geographical Society (with the Institute of British Geographer) carried out the first research project on OpenStreetMap in 2005.

Favorite Map?

"My favourite OSM map is the one that I created for Haiti https://povesham.wordpress.com/2010/01/18/haiti-how-can-vgi-help-comparison-of-openstreetmap-and-google-map-maker. This map demonstrated to me the power of OSM, and the story of its creation is that I created it while physically being in Jerusalem, but the request to create it came from Mikel in the US, the data came from the US (GeoCommons) and Germany (GeoFabrik). The processing was done in a computer in London that I accessed remotely. The volunteers effort that it shows is from many places, and all that to support the relief effort in Haiti."

Interview

Who is Muki and what were you doing before you ever knew about OpenStreetMap?

Okay, so I'm a person who discovered GIS back in '88. And then somewhere in 1991, realized that I can't progress in computing and software without having a degree because I was sitting in all sorts of meetings where people were talking about buffers and registers and things like that. So we're talking about the days of 386, 387, and I was frustrated that I don't know things. So I thought, "Okay, I'm going to university to learn." But because I already knew that I wanted to do GIS, I was going around and figuring out which university will allow me to do that. And the only university that let me do that, in Israel, was the Hebrew University where I could take computer science and geography. And people looked at me as if I'd lost my mind like you take computer science and psychology, you take computer science and business. Nobody in their right mind would take computer science and geography but they let me do that.

And then I continued to do these combinations. So I'm half computer scientist—theoretical computer scientist because that's the nature of the course there—and human geographer with connections to the environment. And that's important for the OpenStreetMap story because by the time I started a PhD at UCL in '97, I was interested in human environment issues to do with GIS. And very quickly I switched into the whole area of public access to environmental information that continued to appear in my research interests. That got me into what's called participatory GIS where you take information and show it to the people.

In 1999, I think, was the first workshop that I ever did. And that's at a time when people couldn't see information. So we received information from the Environment Agency and from the Ordnance Survey and had to bring people to UCL because we were not allowed to take the computers out and we were able to show them the information and I finished. So I've done all sorts of things like that during the PhD. I completed the PhD in 2001 and luckily got a job in the Department of Geomatic Engineering at UCL and continued to be there—if we're talking about before '07, that's what I've continued to do down to 2004 and 2005. And I started doing all kinds of participatory mapping with paper maps and not yet online. I got only online in 2005.

Do you want to talk a little bit about CASA?

You noticed that I haven't, and the reason is—actually, there is an interesting thing is that now—if you can hear my background, it was actually really easy to get into CASA and I was welcomed. So I'm actually one of the first PhD students in CASA in the Centre for Advanced Spatial Analysis. However, I am not completely a match because on one side, I am a system manager and DBA and I also do things like that because that's my background and that's what I love to do.

With regards to the human job of having me, I'm sort of cultural more than technical and I'm very skeptical about, "Hey, look, that's great 3D models" and "Hey, look, this technology is amazing!" I asked "Why are we doing this exactly?" There wasn't always an answer. And all sorts of very clever modeling that I couldn't see a reason for who will need it and why they need it. So although I was in CASA, as the place where I was doing my PhD, I tended to hang out in Geography with the cultural geographers. I feel there's more of their influence on my PhD, even more than CASA. So CASA sort of tolerated me and I've done some things in CASA but ever since, and even during my PhD, I was not completely part of the core family.

When was the first time that you saw or heard about Open Street Map?

Probably at Pecha Kucha in 2005 was it or '04? Something like that. You remember that one? When it was in Westminster University or something?

Okay, so that's when I was getting into the usability of GIS and I was going to get off, but that doesn't mean we were talking about OpenStreet Map. It must be only a few months after you started it.

And what were your first thoughts?

Brilliant! Absolutely. I finally saw a connection between all of my participatory stuff and the potential of participatory technology and what's going on here with open source and the more technical community that was emerging. That got me very, very interested and that's why I thought, "Okay, it's worth following it through and seeing how this project evolves."

So do you want to talk from a research perspective? What you did with OpenStreetMap?

Yes. When you did "a month of OSM", I thought, "Okay, so here's something that I can get access to and then I applied to the Royal Geographical Society to get a bit of money to support bits of work on OSM and also to support Nick Black doing his MSc project. I talked from the start that it would be cool to have something on a mobile phone, and the smart phone of the time was the N95 or N97. I don't remember which of those two models. And the thought was "Oh, can we do some data collection with that stuff?" The answer was "not really." But it was good to try. So that was one aspect of thinking.

The first thought was "How can we make the data collection extremely easy so anyone can just go and collect data with their smart phones?" It turned out it was too early to do that and especially if you think about the function of phones of the time. So that bit evolved into a really interesting conversation. So what was going on is that because of my background and experience, I was linked to the Association for Geographical Information (AGI) and all kinds of established bodies. So I had, for example, a project with Ordinance Survey to suggest to them what should be going on with their data collection, and I mentioned crowdsourcing to them at that time.

There were also other talks with people in the industry. The kind of people that later on presented themselves as supporters, like Steven Feldman. One of the first things that I remember him asking is "How good is this stuff?" And I noticed that again and again and again. The first question that people are asking is "How good is this stuff?" And it had to be no good because those people are—well, they are from the internet and things like that, you know? We don't know who they are really, and you can't get anything good. So that's why I've put myself on the path of saying, "Okay, I'll show you that it's good." Or "Let's ask if it's good." And it was interesting to start researching it.

Now, of course, because of the work I described, the cultural geography aspect, I've got a paper that's now very well-cited about OpenStreetMap and quality. But actually did it because I was interested in something that you said in an interview to "GIS professional" that nobody wanted to map social housing. And I thought, "I want to check if that's true or not." And I needed a quantitative measure that would allow me to compare that. And I thought, "Actually, completeness or quality measure is a good way to check if that statement is correct or not. It turned out to be correct.

But the fact that the paper is actually popular among people who are doing GIS is the factor about quality. But it's not about quality. It's actually about the inequality in the data. So that's why did it. That was the deep motivation to write this paper. And then there were all kinds of other lucky things. So for example, Patrick Weber and I probably should thank you for getting our top-cited paper that John Krumm, I think, of Microsoft Research contacted you at one point and asking, "Oh, can someone write an overview about OpenStreetMap?" And you kind of bounced it to me and I bounced it to Patrick Weber. We worked it in over about a weekend. And that paper is very highly cited as the paper that talks about OpenStreetMap.

So for a non-academic, what would go into that? What went into that paper?

So what went into that paper? I thought that the most interesting thing and it's still the most interesting thing for me about OpenStreetMap that people just don't get it is that the infrastructure is just astonishing. It astonished me at the time, how you can use used servers, all sorts of bits of software, and still provide robust service to 10,000 editors concurrently. And I knew at the time at about the cost. Let's say, the Ordnance Survey server in their Phoenix system that they were trying to do with the 800 editors and that was in the millions.

Can you describe what that is?

So that was a new editing system for all their data sets, to streamline the production of mapping data and being able to edit it and update it. And you sort of know how much it cost, you know the process that they go through and the charges for setting up this type of GIS and here OSM is, something that sits in the corner, with about two and a half used servers that cost I don't know £ 10,000 or something like that with all kinds of bits of software. Actually, the distributed nature of the software internally is fascinating. And it still fascinates me because it also allows organizationally for StreetMap to be kind of lots of kings of the castles--different people controlling different bits and running them. We're talking of something that was considered as a weakness. But at the early stage, it was what enabled OSM to go fast, really fast. And so I thought that that's important in the paper. Some bits about mapping parties and the general process by which the mappers produce but also talking about where it's stored and how it is shared. And also, putting in things like, for example, the aspect that it's not about people just going around with GPS. It is about someone digitizing from an aerial photograph. Plus, running around with a GPS, that makes it happen. So we tried to highlight those things in the paper.

It feels like papers on OSM either do one of two things. It's either cataloguing something that's happening or it's explaining how

something new was built. So this was something highly cited because it was cataloguing OpenStreetMap as a Phenomenon?

I think that's what it is and also somewhat explaining what's going on behind it.

Right. And so, so you talked about Steven Feldman but what about generally in academia. What was the reaction? And how's the reaction changed over the years for OpenStreetMap?

It was always looking at what questions people are asking about OpenStreetMap and how can we look at them and how can I encourage the people that work with me or other people to kind of look and explore that and do things that support a community. So I'm still, for example, trying to convince MSc students to do rigorous analysis of how imports influence the community, you know? Something that the community wants to know.

That for me is a main reason to get engaged with the research because I want to do things that are useful to some community or the wider community of GIS users because we know that even if we'll take—okay, there's a core of a very small group of people that do the technologies, that build stuff, build tools and write the code for that. Then you have the bigger community around them—OSM mappers. But then you have huge a community around them—those are people who want to use the map. And the people that want to use the map, they want to understand certain things about it, so once the quality issue was solved, there were all kinds of other questions.

The next question that was asked was about who will do the updates. So I was also focused on a study to demonstrate how that works. Then you look at—so for example, there is this stupid motivation or incentives? So there was the work of Nama Budhathoki that I've been linked to. And he was looking at motivation and incentives. Then recently, there was some work for the World Bank about trying to convince government officials that it's safe to use this stuff and so on. But that's also part of work that university and academia can do because we look, we are—

Impartial?

Impartial, yeah. And when things don't work, I'm saying that they don't work and when they work, I do say. And you can test it and you can replicate it. So everything is there. And anyone who ever asked me for scripts or data sets, I will get them that.

So can you dig a little bit into quality analysis and exactly how it worked and the results because the people reading this will have no idea.

Yeah. Okay, so the first thing was that surely, it is impossible to get anything accurate because those people are just going around with GPS and GPS is not that accurate. So we compare that OSM data to the Ordnance Survey dataset. And the first thought was finding a method. In '97, Goodchild and Hunter published a paper about describing a comparative method. And it's basically taking two data sets and you know that one of them you trust and the other one you don't trust. If you know for example that that one you trust is expected to be accurate within 20 meters of a position, you can create a buffer of 20 meters on each side of the roads. Then you can calculate how much the untrusted data set falls inside the buffers. This gives you an idea about how close the untrusted data set is to reality because you assume that the buffers around the accurate data set will cover the whole untrusted data set. So we found this method and we thought that's a great method to start with.

And the first thing that I ever compared with this method were the motorways around London. And there were two reasons to start there. First of all, because motorways are big enough. So you would expect then that someone with GPS, if it's completely jumpy it will fall outside the region, but if it's good enough, there is enough space in the buffers for the roads to be captured properly in OpenStreetMap. And also, it's because there were enough journeys on the London roads for the OSM to have captured them. So we have data then to compare with the OS data.

And what surprised us even when we started it is that there was already an 80% overlap. So the next step was to narrow the gap and then check it on streets in general. And again, it became very good. And I stopped doing that when one of the PhD students—a Greek PhD, took the top data set of the Ordnance Survey, called MasterMap, which is collected for each street exactly, with a surveyor, at the scale of 1:2,500. So it's very, very high quality information and the mistakes are not very large in terms of the positional accuracy of this data. It didn't go through what's called generalization. And he showed that, in Central London, OSM is as good as the Ordnance Survey - as the Ordnance Survey actual Master Map. And so OSM can get into the quality of the top data sets that you can throw at it.

Of course you get less good data as you get out of London or go to the suburbs. It deteriorates. The moment you go into Liverpool, you go to the poor part of Liverpool, the quality goes down. In principle, there is nothing stopping OSM from getting the positional accuracy of the Ordnance Survey and that's why— so that's why I kind of said, "Okay, I don't see the point in continuing to ask this question." And then by that time, there was another method that was tested in France. And there was someone who contacted me and had done work in Switzerland. And there was work in the Fraunhofer Institute in Germany. I mean, comparing OSM to Navteq or to TeleAtlas. There was plenty of evidence that OSM can be as good as these data sets—so, you stop.

And so, how was the methodology similar when doing income inequality?

That's actually—the moment you have a measurement of quality – any measurement, it can be used to understand inequality. For example, I got especially in, I think 2007 – 2008 or something like that, one of the reason that I was very interested to be engaged with the community is also to understand the process because if you'll scan the academic publications on OpenStreetMap there are plenty of people who develop things, download the data, just look at the wiki at the time and didn't actually engage with OSM'ers and went to a mapping party and looked at what's happening. And they were assuming all sorts of things. Like, for example, they assumed that most of the data came from GPS tracks.

So they didn't realize that mostly people take information from satellite images and that it's also done with imports. So if you don't understand how the data is made, you just write rubbish in your academic paper and—

And you can point that out for them.

Yeah. I tended to be more polite and not to rub it all the time. But if someone asked me for a talk about it or to point out to them where basically their paper is missing the point. So I was looking at different indicators. For example, the inequality was that once you've done the quality, you can just chop the data according to what's called in the U.K. the output areas. Each output area has got a position in something called the Index of Multiple Deprivation and the rest of it is just a bit of a spreadsheet. So it was just to have a measure—just to have a number that can be compared. Those things were relatively easy and relatively fast. The work on measurement, the quality and determining the quality for each segment was the hard work because the moment you have the two you can start to work against all sorts of social measures.

What would be the headline? Would it be "OpenStreetMap, not as good in poor places?"

That was basically it, I've got to think. I've got it all my blog about why nobody wanted to map social housing estates.

Probably because I'll get stabbed in the face, yes. But what was your answer to the question?

Yeah, I've been plenty of time in housing estates. That's where the participatory mapping work that I did since my PhD was done. You get lovely people anywhere. Just have a cuppa and you get along with anyone.

So if we skip back a little bit, what has your interaction with the OSM community been over that period of time?

I sort of decided to take the position—that's now the social science part in me of doing what's called participant observation—of deciding to be involved in the community but not to be involved deeply. I'm not getting involved with developing the tools or cartography or anything like that. But to be a sort of a critical friend which means working very closely with OSM tools, joining State of the Map and going to meet ups and mapping parties. And when there is any request from OpenStreetMap to try to help you in any way I can, so if it is getting funding or trying to secure the place at UCL or whatever it is that can be done but at the same time, being the person that while understanding it also asks questions back and treating it as a research object that I can ask questions about it and critique.

That, for example, allowed me to do things like going and saying "Hey, it's just 3% women. That's not on, okay?" The sort of things that came out of Nama Budhathoki's research so you can think that you would say, "Oh, that was a biased study. Let's ignore that. Let's forget about this data." Anything like that.

Right. So one interaction I remember was the eye tracking. I think that was you.

Yes. I hadn't done it but I guided or I encouraged Patrick Weber and Kate Jones to do that. And also, because we introduced also the doing of assignments of introduction to our masters program in GIS, it gave us opportunity for bringing in Andy Allan and other people to see how Potlatch actually is being used by students when they're first introduced to OpenStreetMap so that also helps people to kind of improve the usability of it. Also, you can see actually the same approach within the first State of the Map in my talk about what are the barriers that OSM needs to cross in order to be popular.

So why don't we talk about those two things in fact?

I can sit here and say, "I was right, I told you so." You know, when you look at the idea and missing maps, you see that "I told you so" kind of thing. But I'm not that type of person.

Why don't you describe what your first talk was?

Okay, so we need again to roll slightly back. What happened is that when I looked at public access to environmental information, I would look at this participatory GIS in '99 and 2000, that earlier period. What became clearer and clearer to me is that the problem is that all is information was coming in from GIS and nobody—even people with GIS experience—knew how to operate the damn thing, so expecting someone that's from the community center trying to fight the local planning decision to do something, then it was ridiculous. So you need to have better tools, easier to use. And then I got interested in this whole area of human-computer interaction and started developing different research questions about how to make geospatial technologies easier. To make them more useful. I was noticing them in different cases. So when it came to State of the Map, I just put in a paper and suggested explaining what is usability engineering, how you should think about it, why it's important for OpenStreetMap and how I think would be a good thing to do.

When you say you were right, that's what I'm trying to dig down into.

Let's see, that would be 2007, isn't it? So looking at things like for example different tasks that people can do with OpenStreetMap, or getting the data in.

So at the moment, creating the map, I will say that you need the main tags and you need a lot of information and you need to be very computer literate to know what to do with the GPS and to upload it and to use it. And then to think about a new level of map enthusiast with internet connection and time that can be collecting the information quite quickly. An idea is providing a better thing and you can see that people want to have it easier.

Then I also mentioned that you can use a smart phone to collect the data. And I was noting the N95 and the iPhone, actually. And later on there were apps for that. So, for example, what I said there that the needs from the application will be very easy data entry, basic data collection, easy upload and updates and I'm asking, "Can you give us the app for a school teacher?" An application of that is starting to come out. Then getting the data out. So for example, at the moment, there is— I've noticed that there should be a way for GIS users to consume the data and indeed now, through GIS, you got direct consumption from OSM over the web. We've got MapBox and we've got many others and in gadgets also. So, things like that.

Right, so what was your memory of the community like back then?

Kind of very committed, very interested. It was interesting to see the mix between people that are coming with a lot of GIS background and people that are coming from the tech community with less background. So, it's this kind of, you know, reach out to new groups with Ed Parsons appearing in different meetings. But also myself, you know, Chris Perkins, Martin Dodge...

So do you want to talk a little bit in depth about what the goal was, the methodology of eye tracking and what came out of it? I don't know if you were setting up the eye trackers yourself or whatever but it's still interesting, right?

Yeah, so by that time, when Patrick and Kate were doing that, there was already a new generation of eye trackers. The very first generation of eye trackers were a nightmare to operate. Even if you tried to calibrate it, it would go off calibration and you had to throw about five out of ten experiments. But then there were eye trackers that were easier to use, where they're basically sitting at the bottom of the screen. You can tell someone to do a simple task and then you see hotspots of where they are looking at the information and that helped you to analyze how people use the data. So they took different people here and from our MSc course to the laboratory of the UCL Interaction Center, which is the UCL Human-Computer Interaction people, and ran an experiment there to look at how people start using, what happens, what are they looking for, for example for the zoom in/zoom out button which was located in a strange place in the interface. And they demonstrated that they are looking in the wrong place and that helped Andy and Richard to change it to the right place and a few other things of that sort.

I particularly remember that people would look at the left for the save button but that save button was at the top right. And I remember, it. And that surprised me that even with research that shows what people are doing, it was still an uphill struggle to get that kind of change made.

But that's part of our experiment. That's true for anything, you know?

Why'd you say that?

It's the social aspect of the way that everyone danced around everyone else in the change from Potlatch to iD and you need it to have especially, well, from Richard saying that he is fine with all these things so nobody offended him or—and it's not that he didn't mind moving to iD but nobody wanted to offend him in any case that he will be offended and those sorts of things appearing again and again in the history of OpenStreetMap.

So why don't we talk about that? So how would you describe this to someone who's new to the project or new to open, in general?

I don't remember who told me about that but it was something about an anthropologist who had gone into researching mailing lists and other aspects of an open source project and he came back and said, "Oh well, that's just a tribe." So it is a tribe, basically—a group of people that have common goals but also they are very strong-minded and independent and want to protect their independence.

So OpenStreetMap, at the very core of it, got a strong element of both what's sometimes called right wing libertarianism and left wing libertarianism. And the clashes between those sides is what creates all the sparks internally and explains a lot of what's going on. It's very interesting to the onlooker from the outside but it's super-frustrating, I'm sure, from the inside. And it actually tells itself in the way the project evolved over time. So it's a lot of people that are 'kings of the castle' in their small area—that's how I'm also explaining to people the technical diagram of OSM. That a lot of what makes it happen is because of different people. If you told them that they would have to follow what you have told them, it wouldn't have worked.

But if someone can be the person with a log-in to the whole system and the root access and someone else can be the root access to the database, that enables two people to be on the technical side and to do different things and be their own king of the castle. And it's also specific about the people that are doing that. I'm not mentioning names or anything like that. I'm very careful about it but if you look at it, actually each bit of it is allowing this king of the castle and multiple kings of the castle to do their own castles in the space of OpenStreetMap. And all they need to communicate with one another is the API. So they can go on and create their Mapnik or create, you know, Potlatch or whatever. And the market would decide what was good. This doesn't mean that you'd stop using try to stop someone from using things by saying "Sorry, that's not good" for people who are say, stupid enough to try and look at Osmerenderer as a renderer to continue and use it. But if someone wants to do that, they can continue. Even today, I think, isn't it? Is it still there?

I actually don't know. I've no idea.

Hopefully, it's not in–but it might be still running and churning at times but it's important that nobody will stop it. For me, that's the explanation of what's going on in OSM. You'll also find this with the mappers. Actually, I'll be honest. Now, not to be honest, but actually looking at the mapping community, you do see the tension, for example, between the kind of professional community wanted for a specific task and how Ed Parsons once called it the "heroic mappers" of humanitarian OpenStreetMap team. It's fascinating how they collaborate on the database but they are not leading the project actually because of the technical aspects of it. There is a lot of power concentrated among the still hundred or so people that are at the core of OpenStreetMap.

Right, very decentralized.

Decentralized and centralized because those hundred people are still driving what 10,000 people are doing every month?

Right, how do you see this panning out in the future with all these tensions?

Continuing to go on. These projects sometimes go on for a very long time. Linux goes on, yeah? For example, one research part that I really want to do sometime is to analyze in a lot of detail of what was going on during the license change because socially, an organizational change like this is a really interesting moment and what was going on during the license change is that yes, there was a fall off. It didn't last and it wasn't that disastrous. Okay, there was Australia. A sub-case but, you know, that's not a big thing. On the other hand, it's still remembered across the community as a traumatic event, which is why people don't want to reopen it. "Oh, I don't like, share-alike." You know, that kind of thing.

Yeah, because it would be very beneficial to their business model if there wasn't share alike.

Yeah and that's what people think but nobody's saying it. Well, some people are saying it.

Yup, I do.

Okay.

So your characterization of left wing—I think you said left wing libertarian and right wing libertarian. I think that resonates with me a lot because I think, when I started the project, I was left wing libertarian and now I'm right wing libertarian. And that explains an awful lot over the years.

But as you were saying, and I don't know if you're still saying that that's why an open source project really needs Germans.

Yes.

Remember that line?

Yup.

The point is that the Germans are all in the left wing and if you kind of analyze it, What symbolizes it more than anything is that Frederik and Jochen can't work together—you know, it's not that they can't work. Jochen is very talented but he can't work with anyone. They are very talented, very capable on many levels but do you see them running an organization with 20 people? Okay, two, but not 22.

Yup. It's funny because I have a friend who actually hired them.

Yeah but at the same time their positionality in terms of what they believe in is fairly within the left wing libertarian.

Right, it's interesting because I had a friend who hired them. And they said, "Frederik was splitting up with Jochen because the company was getting too big." It was two people. Anyway. Having worked at Microsoft with a 100,000 people, it's, you know, different scales. So what's been the thing that's sort of most surprised you about the project?

Actually, the most surprising thing was that I thought the fight on convincing the mainstream GI science people to get it in, to accept it, would be much, much longer and much tougher. And it was over within—what is it, three years or four years? You know, it wasn't even completed in the U.K. or in Germany to the degree that it is now. I thought that it would take something like five years or something like that but it was all over by three years. So that, for me, in the history of OSM was the most surprising.

Can you talk about that process and how you noticed the change?

So what I noticed is that I only started talking about OSM and all this time I'm getting the same crap that I'm currently getting about Citizen Science, "You can't get good quality data off of it." And then when people are convinced and I showed them all the slides about good quality data, they immediately switched into "Yes, but they won't continue to collect it and do it for a short time and they'll give up." And when you show them that people continue to collect it over time, then they'll go on "But we don't understand the motivations and incentives and they're all volunteers." And the answer is, "You don't care about—why do you care about motivation and incentives? Do you care about the motivation and incentives of every surveyor in the OS?" But they'll throw you all sorts of ridiculous stones at their own thing and I'm puzzled. And we get it a lot in the citizen science area that I'm in it now, but those questions very rapidly disappeared. So actually people accept it and you start seeing people presenting it.

Again, a big surprise for me was—I always thought that the people that would have the biggest resistance to it would be surveyors—so land surveying, land registration. I thought, "Okay, everyone else, on the base map, people will accept OpenStreetMap but when it comes to surveying, they will say that because there is a lot of money in being a surveyor." In the States, it is the same way. A few years ago, surveyors in the States tried to stop anyone else from creating a map for one thing. They claimed that only a surveyor could produce a map or something ridiculous like that so I was sure that they would resist it. And then they came out with a conclusion, "Nope, there aren't enough surveyors in the world. We can't be everywhere. We'll need to figure out how to do data collection by the crowd." So all these things happened by people realizing that this works. There is enough evidence. There was indeed not just us but lots of other people showing that the quality is good. So there was no reason to stop it.

I think it's also the realization that we want big money. I think that that also helped. So after 2008, when everyone realized the money stopped, you know, they want to be more and more fascinating in all sorts of things. That also convinced them because before then it was "Oh, like if only the world invested in huge amount of surveying. If only the world invested in—"

There was still someone actually, Hernando de Soto, who goes around the world claiming that the solution for all the problem in the world is to create a cadastre. That way, the market will be supported by the State and forget that the whole thing is relying on all sorts of many other things but that's what this guy means. He's even got a book, *Mysteries of Capital*, that makes these claims about registration. So the claim is that if the World Bank will invest enough in land registration, all the problems in the world will be solved—something like that. More and more there's a realization of crowd sourcing, with people like Robin McLaren.

Do you have any thoughts on the humanitarian aspect?

Yeah. Amazing, so that's another fun story that I have because I have written about the quality data. When Haiti earthquake occurred, I was on a family visit in Jerusalem. My computer at UCL was running and I would just, you know, I'm doing some processing of OSM data or one of those things. The computer was running in the background and Mikel wrote me an e-mail about, "Oh, Google was saying that our data is rubbish. Can you compare the OSM data to Google to show how good we are at collecting the data?" I was in touch with Mikel on other things. So I could log in from my hotel in Jerusalem to my computer in London, download the data that OSM had. I had already all the algorithms ready to compare the quality of what was in the U.N. data to Google's data. And then the blog post that I wrote about was one that got the highest number of hits on my blog and it basically helped to demonstrate and Mikel and other people that OSM was doing the right thing and collecting the right data.

What were the results?

The result was that the Google interestingly had some data in Haiti. The data they were collecting even in the places where there wasn't the earthquake. If you look at it, they actually had previous data that was collected close to what is it? Between Haiti and Dominican Republic. You can see it on the blog post. It's really interesting stuff. But so I've stayed on the humanitarian OSM team from the start and I'm a great supporter of them. It's amazing that it's actually a place where you can see that people are very happy to contribute. It would be interesting to check if that's now a major source now of people learning OSM.

Right, do you mind digging in a little more into the results? You know, what the blog post says? I'm just putting myself in the shoes of a reader who doesn't know.

So my first post is from January 18th, and the earthquake was on the 10th I think. And by the 18th I was comparing OSM and Google Map Maker. And I'm kind of looking at it and doing basic comparisons of the data that existed in OpenStreetMap and the amount of data that existed in Google. And around Port-Au-Prince, you will see much more data in OSM

It's been just over 10 years of OpenStreetMap. How else have other things—especially things in academia changed? We've touched on that the iPhone exists now and it didn't then, for example. But presumably there are other things also.

Yes, Big Data became a big topic in the area of GIS, in the area called Geographical Information Science. What OpenStreetMap is doing became known as volunteered geographical information (VGI). This concept was invented in 2007. So it became a legitimate area of research, which is very important and kind of led people to download the data and do things with it.

The way universities and this type of public research works is that when you have a topic and someone sets a research agenda, and if there are academic papers published about it, then people apply for funding that then allow other people—PhD students and researchers—to work on this topic and generate more stuff. And sometimes an area becomes a major area and continues to evolve but sometimes it's just for a period of time.

So VGI and this whole notion of crowdsourced geographical information—in the States, it's morphed into something called CyberGIS. Horrible name, even worse than VGI. And in Canada and in other places, it's now a called GeoWeb which is actually a better term for the whole thing that's happening on the web. And I belong to another morphing—a specific area called citizen science which actually has been going on longer and more widely with communities that include people in ecology and in bird watching and in physics.

In terms of the students, they expect to know about crowdsourcing and this information. So you have official funding and research. And there continues to be interest in geographical information but you see the research now also in computer science. So there is now officially an area called GeoHCI. You know, if I talked like that in '91, people thought it was mad to talk about computer science and geography, now you've got a sub-sub-discipline that is dedicated to it and a thing like that.

So I'm curious also, has anything changed in terms of a decade ago, if you wanted to do research and you needed a map, it was sort of hard and expensive. But now, if you want a map, it's sort of it's just easily available. And how has that impacted anything that you've seen?

That's not true.

That's not true?

No, because you could get anything you wanted if you were a researcher, even going back to '88. And you just made contact through our university mail and they would send you the pack.

Okay, well it impacted me as an undergraduate. I can tell you that.

Absolutely. And that's the reason you've made OSM. So while you were hanging around CASA, in the VR Center, you had access to all the data, so you would come to say me or you know, Paul or anyone else and ask for data. Anyone would kind of get into it and get you something or whatever but the moment you stop then suddenly you're off. But one of the reasons why I came to the U.K. to do my PhD was because I knew that there was such a richness of data. It's astonishing. I still remember how astonishing it is because when I was doing my Master's degree. Although I was in the university, I had to negotiate for three months with the Israel Mapping Agency to get any scrap of data and from someone who was keeping all the ecological records to ever get me the data. And here I am sitting around and clicking on some buttons, getting the data and visualizing it and doing fun stuff with it.

Do you think that was because Israel is surrounded by enemies or it's something else?

It's the same problem in Malta to this day. It's a problem in many places. So for a lot of places, OSM really changed things. You can look at how many papers there are in the States because of the availability of data. They rely on TIGER. They didn't rely on the improved product. You know for yourself that the data is rubbish when it comes in its basic form but that's what researchers could access. They couldn't access the improved product because for that you would need to have money or to get permission from the company and then it would be better for what you are doing. So research is always kind of shifting according to the availability of data. I think that nowadays, that is an important thing. Convenience is really, really important. So people – researchers trying to do something, a lot of times they are lazy in different ways. So they don't want to take data and start messing about it, if you can just click on your browser and get your data from a site which is easier and it's therefore used much more. So I mentioned when, kind of recently, I tried to analyze the type of studies that are done with OSM, you'll find the biggest group of studies is actually by people who just find it easier. It's easier in terms of copyright. It's easier in terms of use. That's what you do.

Okay, so what is the thing that you'd want to most improve about OpenStreetMap?

I don't think that the interaction has improved significantly over the years. The ability of people to contribute in many ways. There are now people that are serious about it. I'm noticing in each visit there are more people maintaining the wiki and improving that data, that's all very nice to see.

Is there anything else you want to talk about that we've missed?

I'd like to hear a bit from you about what I've told you about—the research, the aspect, the way it looked to you. Does it matter, for example, among the people that you are interacting with that there is economic research about it or something?

I think it matters a lot. I wish that people in the community cared a lot more and that they would listen. I remember, it was me that submitted the patch that moved the save button from the right to the left in Potlatch as an example. And I just remember just how painful it was just getting Richard to see that this was a problem. And you had slides or someone did slides of one of the state maps. It's like it was in black and white in like six feet across. When people click save, they look here and the button's here, right? And it's never been a problem for me to accept feedback and to fix things. It annoys me so much that most people in the universe are just completely incapable of taking feedback like that. And they all cling on to some sort of posthoc rationalization

of why they've decided something, you know, rather than re-visiting their assumptions. But I think it's super valuable.

To your question, yes it's useful and yes it's valuable. People bring it up all the time, you know, because they seek a structure that gives it a stamp of approval, right? And that's what you're doing with those papers that say, "Hey, it's not terrible."

That's exactly when I knew that there's power in these stupid walls. I think it's ridiculous. You know, it's walls like any other office. The fact that it's got UCL on the front. Like, we both know that even 15 years ago nobody knew what UCL is. Just been extremely successful PR machines working somewhere mysterious and making it into a known place, seriously. Somehow I'm not sorry, but it's an opportunity that I do have so I have to use it. So actually, this is how it works. Very, very personally, for me, it worked well. So I'm now involved outside here with a group of 20 people that work with me. And all the other nice things that come with it. And people stopped asking me about stupid agent-based models. I can live without that. That's an example of a complete waste of time that universities can do.

So I'm curious—

It is useful for what? Nice games but you know? It's not doing anything.

So what were your thoughts on Space Syntax?

Fantastic, it's doing really, really nice things now. We're actually using OSM and Space Syntax in Nairobi to study a slum. We are in the slum. Here's an example of mapping that no government will do. And the reason for that is very simple. If you're like myself, you want your map to give legitimacy to people who live there and we're doing this project with medics about the transfer of disease from animals to human through street vendors in some slums in Nairobi. And there is no other data to take other than the OSM, so we use OSM and we use balloon mapping. What we do with that is figuring out where the switch then goes off and we use Space Syntax to understand why they choose the location that they choose. So for that type it's really, really cool to use it. And I've been fully operating with Space Syntax for the past 10 years after I left CASA. CASA was kind of in conflict with Space Syntax. I'm not completely sure why, so it is useful in some things, though like anything else, not to get over the top.

And I'm so curious. Do you have any good examples of these of not mapping it, like you literally have a map and there'd be like a shanty town of a half million people and it wouldn't be on a map?

Yeah. There are examples from India and from places like Nairobi—You can talk with Mikel about starting Map Kibera.

Henk Hoff

Bio

Henk Hoff is a freelance IT consultant. He organised the State of the Map events in Amsterdam (2009), Girona (2010), Denver (2011), Tokyo (2012), Birmingham (2013) and Buenos Aires (2014). He is member of the Board of Directors of the OpenStreetMap Foundation since 2008.

Favorite Map?

"I've actually got two favorite places on the map: The City of Assen, the Netherlands. One of the few places in the Netherlands that was touched by the AND import. And which I helped complete in 2007. And Pyongyang, North Korea. I started the map of North Korea in 2008. Ever since I visited this country several years before, I'm highly interested in what's going on in this country."

Interview

Why don't we start with who is Henk and what were you doing before you found OSM?

Well, the time that I got interested in OSM, I'm pretty sure because I've just looked it up, it was April 2007. At that time, I'd just changed jobs. At the beginning of that year, I ended my regular job and I went freelance. I was looking around to see if there was anything I could maybe specialize in and/or make something a little bit my kind of thing, where I could excel as a freelance consultant.

At that time Google Maps picked up. I was looking at that and I just saw the potential of putting things on a map. Linking it to location because everything has a location, and being able to put something on the map was fascinating to me. Google Maps was one thing, so I was also looking around to other kinds of mapping products.

I stumbled across OpenStreetMap. It was very limited at that time because if I looked at my own city, it was basically only the freeway that was passing through the city, and the rest was just blank. There was nothing there. But the whole thing about just playing around, drawing your own streets and creating a map was an interesting thing of like, "Okay, now, I can play with data and not just put pointers on the map. I can really play with data."

With OpenStreetMap I could start to play around with creating maps. And when I had something interesting, I could just buy a full set of data and do something fun with it. This was a great opportunity to do something without spending any money but just spend a lot of time and play around and try to see if things are working. And it was—how do you say that—addictive. All that virgin land for me to explore.

Shortly after I joined, I found another person in my hometown who was also very active in creating his own map. We joined forces and tried to map our whole city and we achieved that in a couple of months. It also helped that Richard van der Weerd already had lots of data in his proprietary format; which he exported to OSM. That really was a fun thing to do and gave us all the idea of the potential of OpenStreetMap.

And from then on, I started to explore a little bit more than just my city. I was getting more involved in OpenStreetMap with mapping, and also trying to evangelize and talk about OpenStreetMap to organizations. That was basically my beginning with OpenStreetMap. Actually, within the year, lots of exciting stuff had happened. Starting in April, for me, somewhere in June, that year, together with Richard van der Weerd we had the whole city mapped. And during the summer AND announced that it would donate their map of the Netherlands to OpenStreetMap. Which was another major thing. It was an exciting ride to work with OpenStreetMap.

What was your involvement with the community at the time?

The community, at the time, in the Netherlands was very small. I think was about ten persons who were active. The interesting thing about the Netherlands was around that time, they had a grant from the Digital Pioneers. It's an organization linked to the Ministry of Economic Affairs to stimulate digital projects. OpenStreetMap got a grant of – I forgot how much but a couple of thousand Euros—to start building a community. With that money, we could do lots of mapping parties and all kinds of other stuff. We actually also bought GPS devices with it. It was at the time a very small community. But we were really working closely together—dedicated to trying to start up something new. Most of the meetups we had were basically done in Amsterdam.

Can you describe what those meetups were like?

One of the people there, Zoran Kovačević, owned a software development company in Amsterdam. We just could meet up at his office with a group of between five and ten. It was, in the beginning, more explaining how OpenStreetMap works and so how you could edit. All you basically had, at that time, was the JOSM editor, no aerial imagery, no smartphones, just GPS devices. You really had to have those Garmin devices or some data loggers that you could use while walking around.

We just tried to map Amsterdam, dividing it into sections, and we all split up and walked or cycled around. It was pretty difficult in the beginning especially if you were in downtown because then you had those narrow alleys and that was very difficult to get a very good fix on your GPS. And again, remember, no aerial imagery so you only needed to have something like, "Okay, yeah. So this trace you got from the GPS – here it's getting wonky and it's not really correct." And it's trying to decipher the trace you got from the GPS whether it was an okay trace or not.

In the beginning, we tried to establish what the purpose of the mapping party was. Was it just going to a place that wasn't mapped tracing buildings? Or go there to check whether it was complete, whether alleys or streets were missing. And mapping just the basic amenities—bars, restaurants. Well, I think, that was basically it at the time. Just getting out there to map for an hour, and getting back try to—well, yeah, either we're having pizzas or we went our ways. It was just a bunch of geeky people working together on creating a map, and we were very enthusiastic.

I still have, somewhere, a printed map. One of the first maps that was done I think right after the AND import. We said, "Hey, we have a whole country map." So you can display it. "This is it."

It's very difficult to say but lots of things happened and it was really quick. Also, setting up our own map server was a little bit difficult. You really needed to know what you were doing. It was not very well documented on the Wiki. It was just really basic stuff that you needed to do and you still had that – it's not so much Mapnik, you also had that SVG thing. Gosh, what was it called?

Osmarender.

Osmarender, yeah. I know from my personal experience—that sometimes Osmarender and sometimes Mapnik, and also knowing that Mapnik at that time was just basically you could draw a line there. It couldn't do all kinds of nice rounding and nice color effects and that kind of thing. No, you were just drawing a straight line from A to B and the text was just displayed at a certain place. And it was sometimes horrible looking but at least it was a map and you could see what you were doing and that you could actually create something from blank to actually have something on a map. It was fun.

Gosh, I can't think of how many meet ups we had at the time. I remember that before that, I hadn't been to Amsterdam that often. And I think in 2007, I doubled my visits to Amsterdam in my whole life just in that one year. Okay, the Netherlands is not really that big a country, but I basically live on the opposite side of the country, so it's about a two-hour drive or trip to get to Amsterdam, but still. Well, it's a doable trip. I think that was one of the things—I remember at certain points that I thought, "Oh gosh, I'm in Amsterdam again." And at some point, I said, "Ey, I know this street. I've been here before." To just get a feeling that you're in a city and that you somehow get a little bit acquainted to the city itself where well, yeah, it's just basically a very foreign terrain for you.

At some point in 2007 or 2008, we were invited to a discussion, and it was about the EU INSPIRE directive. A guideline to open up data within the EU. The main participants were OpenStreetMap and Falk Plan. The latter being a company that sells paper maps.

There were accusing us of stealing data because OSM was mapping structures which couldn't be done by just by manual survey. There were small lakes on the map which were not detectable on the coarse grained aerial imagery. "How can you put the lake on there? You must have copied that from paper maps" When I replied with, "You can walk around it with a GPS" (which I'd actually done for a small lake). Falk Plan: "Haha, no, no, no, serious. I'm asking you a very serious question." Me: "I am giving you a serious answer. This is what you can do."

That was an interesting thing to see how the establishment was thinking in a very, very strict or very narrow perspective of "this is the way things work and there's no way other people can do it better, or can actually do it. This is the way things work and/or how it should work." So having these kinds of discussions, trying to make them aware that there is some quality in having non-professionals or volunteers trying to do the same stuff, and that might be good enough for certain purposes. So that was one of the first times I was in a discussion with a commercial company. It was kind of close to badgering: "obviously, they can't do it. They're frauds. They're stealing data. They're geo-terrorists," although that was not really the word that he was actually saying but it was close to that.

And that was somewhere early in 2008. We're now in 2014. So six years later, it's a no-brainer to people that OpenStreetMap is a very good and reliable source. Just in six years, coming from a place where the industry is completely slashing you and it's totally impossible, "You're whatever." And now, they're embracing OpenStreetMap. It was like, "well, this is one of the biggest data sources available and it's also rich in quality, et cetera." It's still kind of mind blowing in how short a time that could've occurred.

Back to 2007, somewhere in July—around the first State of the Map actually—the guys from AND announced their sponsorship of OpenStreetMap. It's mostly that they donated a map they had. It turned out to be quite an old version but nonetheless we were happy with it, of their maps from the Netherlands, China and India. Sometime in September, we were very busy trying to see, "how are we going to import that data?" because especially in the Netherlands, it happened very quickly. Having that data donated, there was, in the beginning, a kind of a discussion, "Should we actually use it? Should we actually import it? Or should we actually do something with it? Or what should we do with it?" So we didn't immediately decide to say, "Hey, let's import that stuff." Because at first some of the people had a look at it and asked how the quality of the data was. because they didn't want to have junk in the map. Also sentiments like "It's a commercial company and they're giving away data. That might be just a Trojan horse or something like that. Or if we import it, what is going to happen?" It was not just

"Hooray, happy" - that kind of thing. So we're just kind of questioning their intentions. "Is there something that's biting us in the ass or whatever? Is that actually good quality data?"

We had several debates about that and also about the process of how to import it. And at a certain point, we basically decided that the quality of the data, in most places or in by far the most places of the Netherlands, was better than we had at that time except for certain places. So Assen (my hometown) obviously was one of the places where we were significantly better than the AND data we got. There were some other places that were also better mapped than the AND data.

So we came up with an idea of "Look, hey let's sweep. Well, let's delete all the main roads or the road categories we had gotten from AND." So, basically let's say the unclassified until motorway kind of things. So let's delete the whole thing and import AND data except for certain areas where we think our data's better. And we needed to draw polygons.

We had lots of discussion with the community on whether "do you want to keep your data?" We actually e-mailed I think everybody who had contributed at that time to the map in the Netherlands. I still think that we had a discussion there. "Can we just e-mail everybody? Is it not spam or whatever?" But at the end, we did e-mail everybody with a plan. "This is what we're going to do. So we're going to delete these kinds of data. If you do feel that your area is very important and that you do want to keep it, this is the way to do it. Draw a polygon around it with certain tags so we won't delete it. We'll keep it in there. But then also, your responsibility is then to reconnect it with the imported data from AND again. So that was, I think, a very good process on how to import. So it was not just an import—we had discussions. We contacted everybody again so we could also get some attention to some early adopters who'd just gone away, we could re-connect with them like, "Hey, this is what we're doing." And at the end, we did import the whole stuff.

I think, also at the time, we had contact with—I don't know who we were talking to in the U.K., but it was also the kind of thing like, "Okay, if we're going to import this whole thing, we need to warn people because of lots of data is coming up. So is the API going to hold it? Are we able to get these kinds of things? Then re-generating the map was going to be very, very slow. I think we didn't have tile caching at that time. It was just if you looked at the map or it was just kind of an instant generation of the map. So just knowing that you'd have so much more data, it became clear that you also need to have the infrastructure on a next step because it's not just a certain amount of roads there.

But I think for us, for the Netherlands, the most logical question then probably is going to be "How did it affect the community in the Netherlands?" I would basically say the community in the Netherlands was really, really in a kind of infancy status. It had not really grown. It was just people trying to do something but it's not really a very big community. And some people there were just interested in OpenStreetMap because you could just create new stuff. Or you could create new things on the map.

With the AND on board, it was basically, "Well, hey, your standard road infrastructure is just done." So every road was in there, so what's next? There were several people who just felt like, "Okay, there's nothing else more to do. It's not interesting so goodbye and I'm going to look for other ventures." I think, at the time, it somehow—I wouldn't say diminished the Dutch community—but it it really hurt the Dutch community because there were several people who were very, very dedicated and interested in OpenStreetMap. They thought, "Well, hey it's not very interesting anymore because the map is done." The community was not big enough at that time, so I got a bit hit.

So if you look at other kind of things, so the U.K. was going very well. Germany was going very well too. Both a very strong community, because they basically created their own map. And the Netherlands is basically a major import. So the whole thing about ownership of your map wasn't really created very well in the Netherlands. In the U.K. you had very more strong people saying, "Well, this is our map because we've put all kinds of blood, sweat and tears in it." Same kind of thing with Germany who was also very, very keen at that time or very, very dedicated to map at the time. And so it had an effect on the community. I think because everywhere else, there was still a kind of a community there.

I think it's more that at the time when you had that map, so the AND import was done, there were lots of questions outside the Netherlands, "so what are you going to do now since you're done?" So the whole principle at the time—the mindset was "we're just an OpenStreetMap." So if you have every street in there, you're done.

And at the time, lots of people still kind of think "Well, mapping the streets, roads and that kind of thing." We were now more in the phase of "Hey, we now need to switch into a maintenance mode. Or that we need to add other stuff." It was interesting I think that since the AND import didn't have all the cycle networks, more people could focus on cycle ways since the Netherlands is a very popular cycle country. We have quite a lot of them.

And at some point, the idea also, I think popped up, "Hey, you're actually never done. There's always something new." So if you put your bar to "we want to have every street mapped." Yeah, and so when you reach that point, there's a new goal you can set. There's always something new. There's always something else to map, or to improve, or to whatever you want to do with it. I think the idea of—I wouldn't say think out of the box—but just thinking a little bit broader than what you were doing at the time created that more people - and especially people who were not so much dedicated to car navigation but were more into cycling or walking but also doing lots of applications with the data. Trying to create applications using OpenStreetMap data like routing.

So in the Netherlands at the time, we didn't need to focus so much more on mapping itself or just getting a map. It was more like getting to the next level and getting more details. Getting more details in the map or getting other things that you can map. So like I said, cycle ways, walking foot ways and that kind of stuff to try and make the map more detailed. And also, fixing things because in the Netherlands, I think, about 10% of the road network is changing every year so there is lots of maintenance that needs to be done. I think people were kind of catching on to that.

And the moment that OpenStreetMap caught on by being used in all kinds of ways, there were lots more people trying to see how you could improve the map itself. So the Netherlands is coming from a different perspective than most other countries because it's basically the country of the imports. Yeah there are quite a lot of imports in the map in the Netherlands, not all of them good. So we've learned about these kind of things, and I think the AND import was one of the better ones.

How did you go from this sort of involvement to having organizational involvement with the Foundation?

That's an interesting question. I had been to the first State of the Map – Manchester, I think it was. I am a person who I would say likes to organize, likes to contribute also in of soft kind of thing, not so much in diehard programming but also in trying to get involved in setting up an organization, running things, presenting stuff and that sort of thing. And because I've been involved with all kinds of other organizations and I was someone who would actually plug into who's interested in working on organizational stuff within an organization. And OpenStreetMap was kind of fascinating and I was kind of attracted to it in trying to see how can we move this limit forward.

And if I look at the Foundation, I think it was in the very early years of the Foundation at the time. It was mainly a U.K.-based organization. Every board member or every person who's mainly involved within the OpenStreetMap projects was basically based in the U.K. and I thought at the time it would also be good to have other people within the organization outside the U.K. to broaden the view of the organization. And so, at some point, I ran for member of the board. Well, yeah, we didn't need to have that many votes to get actually a seat on the board. I think the first time, I probably got about 19 votes and I got elected. That was kind of a weird kind thing. And it was not like we really had a very, very big organization. It was more like, "Well, yeah, it's one of those kinds of things that you need to have to try and organize and try to get things done."

But it was kind of interesting, for me at least, to be the first person outside the U.K. and also the first non-native English speaker on the board. And that was, at the beginning, sometimes a little bit difficult because I wasn't really used to having meetings on the phone, with a phone call and to just be in a meeting with people all talking. All being native English speakers, they tended to speak fast. Also, the meeting culture was very U.K. That was weird to me. Well, having minutes is okay but then you propose minutes? Like "Okay, is that the way you're doing it?" So the whole thing about having meetings was different in the U.K. than it was in the Netherlands where I got used how you did run meetings.

So the beginning was very difficult for me to try to keep up and listen to Englishmen speaking very fast in English and trying to comprehend what they're talking about. But it was a good thing to be there and try to work things and try to build an organization. And I think the first year for me was trying to figure out what my role could be or basically also "What is the Foundation actually?" And it was still the kind of the smaller one. It was, at the time, I think also focused on trying to build the community.

So again, the first State of the Map—that was where I wasn't really involved in it or I was just there. I think it was the first or one of the first, I don't know. It was at an OpenStreetMap meeting outside the Netherlands to hear and see people - other, I would say, geekier people talking about maps. That was kind of weird.

So I think for me OpenStreetMap just came along at the right moment. I was looking for something new to work on. Like I said in the beginning, when I started with OpenStreetMap, I was looking around because I just started freelancing and looking around for something interesting. And I think, at some point, I decided that OpenStreetMap was that interesting kind of thing. And for me personally to just get a little bit more involved in organizational kind of stuff was the next logical step for me. Yeah, gosh, why do you get involved? Gosh, I think that's just might me be one other high philosophical questions of life. "Why do you get involved in things?" I don't know.

Well, what do you think you achieved as a board member of the Foundation?

There might be several kinds of things. One of the things that I think really kind of sticks with me is State of the Map. So the first two series of SotMs were Manchester and Limerick. I was just there as a participant. From then on, I was involved with the organization trying to make State of the Map bigger but also going to new places, building the community. From Amsterdam to Spain (Girona), the US (Denver), Japan, UK and Argentina (in 2014).

So the first one, I got involved with was in Amsterdam and that was because I did a bid for Amsterdam. So, again, I've been to the first one. I've been to the second one at Limerick. Then I thought, "Hey, the next one should be in the Netherlands because we had an import done." We were one of the leading countries at the time if you look at how the coverage of the map was and trying to get it outside the English-speaking world. I started the bid and got other people interested in joining. Martijn van Exel, Floris Looijesteijn and the guys from AND helped out and some other people.

Amsterdam was the first time that we had a three-day conference because we also felt, "We need to reach out not only to the community but also try to reach out and so the idea of a business day or commercial day popped up—the Friday, the extra day, in trying to showcase OpenStreetMap." So, in Amsterdam, it was mainly about showcasing OpenStreetMap - what could be done with data itself. And how can you use OpenStreetMap in all kinds of applications, situations, et cetera. It was an attempt. I'm not quite sure if we completely succeeded in the original idea of the commercial or the business day was but having that, it did try to do a little bit more. Not only that we were kind of an introvert group of people only looking to each other like in this level of thinking of "Ooh, aah, what a really cool project." But also, trying to make each other know that OpenStreetMap is a little bit more than just a fun project to map on and that kind of thing but that it can also be used and we would try to do all kinds of interesting stuff there.

I think it was a very, very good conference in the end. And we also got very good feedback from the organizer and from the City of Amsterdam. They were kind of in awe because they normally had these kind of typically, I would say, bureaucratic events where there's something and this was just a community of geeks who were just running around with laptops and that kind of thing.

Oh yeah, the "internet in the box". So, you got all kinds of weird things like internet in the box. So they had a problem with their internet connection. They just had a very, very small internet connection. And because they were part of the City of Amsterdam, they were just on the internet connection of the City of Amsterdam and they were not used to having so many devices there or that kind of thing. So we basically couldn't use their internet system because the minute somebody would do something, it would just completely go blank and then probably lots of the other departments in that building would also not have internet connection. So that was no-no. Yeah, we had two or three devices of those internet-in-a-box kind of thing. It was not a 4G but it was another thing. We were lucky to have it. In Amsterdam, there was a 3G+ kind of internet connection so we had a couple of those boxes somewhere, standing around in the room. I think it was more like a suitcase. It actually said on the box, "Internet in a box." And I really kind of remember that people put pictures online of this internet-in-a-box thing. And that internet connection was just horrible. It didn't work or it wasn't stable or lots of people downloaded complete planet files during the conference so that was one of those kind of things like, "Okay, yeah, we definitely need to do something different."

Yeah, gosh, there's so much to tell about Amsterdam actually. Maybe one of the things that went on for several years after is that famous positive or negative auction. It started in Amsterdam. Well, for those people who don't know about this auction. Well, the story will explain kind of it. During the last day, Sunday, I think it was me who just asked, "Okay, so we're near the end. And then gosh, we have all those banners. We have all those other kind of stuff. How do you get rid of it? Where are we going to put it?" And, at some point I said, "Oh well, we can just auction it off and if we can get a couple of Euros from it, well, we are rid of the 'junk' and so we don't have to dispose of it." And we got just a slightly little bit of cash in it. And I asked several people, "Hey, this is an idea," I think, in the beginning. Well, this was definitely in the organizing team and I think they said basically, "If you think it's a nice idea just go with it." It was not I think really greeted with a great "Yeah!" It was just, "Well, let's see. Everything can go."

And for some reason, it turned out to be a very good success, I would say. I think, actually even people like the IRC worker were bidding on things. And it was one of those things that started off as a, "Well, yeah, we can do things." And somehow it was so successful that, in years after, there was kind of a cry, "Hey, yeah, we need to have an auction." So it basically started off in Amsterdam—this auction thing, as "Well, yeah, those banners we need to get rid of it. We could stash is at our places but it's just there and if we can make people happy with it." So we've actually had some of those banners go to, I think, Mikel Maron bought one of them and he went with them to, I think, it was Palestine or somewhere in Africa. So that was kind of an interesting kind of thing. So that's—yeah, Amsterdam. What else can I say about Amsterdam?

So what was the difference between Manchester and Amsterdam?

Manchester, was—well, it was just a community thing and it was just a start. I think it was also the idea of "Hey, we need to get people together. Why not do a conference and talk to each other?" Limerick was more or less the same. It was a good kind of thing but it was basically trying to tell each other how great OpenStreetMap is, which is a good thing because you actually meet people, but you're talking to the in-crowd. Amsterdam was the first time that we also tried to say "Hey, it's not only looking at ourselves. We also need to look outside. We need to try to get the external world inside and get the external world in to show off what OpenStreetMap is about." I think that's the main difference, the main thing that made Amsterdam different from Manchester and Limerick.

So yeah, I think that's the main thing and it being in mainland Europe, so lots more people could actually go there without needing to fly. And also, in the number of sign ups, I think Manchester was around 100. Limerick was around 100 or a little bit more. Amsterdam was close to 200 because I think 200 was the limit we had because we were not allowed to have more people in the venue. So it was one of those of successes in that more people went to it. You didn't have only mappers there, you had also people from other companies just to kind of see. "Hey, what is this OpenStreetMap thing?" They were pretty much in disguise or they didn't really say where they were from but at some point you could kind of know. Well, if you're in the Netherlands, you can bet that there were people from TomTom there and basically the other companies. So I think Amsterdam was when we really began focusing on the usage of OpenStreetMap and how could we showcase Open Street Map to the greater outside world.

Amsterdam had lots of good reviews on the first day for the food. That's also one of those interesting things about these kinds of conferences because you had all these kind of cultural differences. It has nothing to do with maps itself but how you organize things. So the Friday, we had, this special kind of thing. So we had a typically duck kind of thing. So we had a herring stand there with traditional costumes and that kind of thing.

Are there any highlights from other conferences you'd like to talk about?

The interesting problem we had with Girona was the budget, because it was a little bit more expensive than we initially thought. So at some point, with every person signing up, we were actually losing money because the catering was more expensive than what a person actually paid for his/her ticket. And at some point, we did have a crisis meeting like, "Oh God, what do we need to do? This is going to be completely out of control." And we fixed it by basically ditching the in-house caterer because we were at a conference center in Girona. If you were there, you basically needed to do the catering by the in-house caterer. You couldn't go to another caterer or you had to buy them out. And that's one of the problems with conferences. If you have a venue, the venue might be okay but then you are stuck with the caterer that's in the venue.

We got around it because we were lucky. And so, we had a discussion with the venue owner on "So how is this thing actually working? How can we get around it?" And he said, "Well, yeah, so this is the exclusivity contract that the venue had with that caterer. So in a line or so many meters around the building, they had exclusive rights for catering." And at some point that was for us the idea of "Hey, wait a minute. So if we go outside and if we go far enough outside, we're not linked to the caterer anymore?" And that was our escape, because there were several other caterers inside Girona that were really interested in helping us out for a very, very much lower price.

The result was that during Saturday and the Sunday, we had the catering outside. So you had to walk outside for, I think, about five minutes to actually go to a grassy field in a park. The break was having a picnic under some trees and a nice park. And well, Girona, Spain is known for its good weather—it was just basically a no-brainer that it would work out. Having the coffee break, tea break or lunch break outside and people needed to get outside the building, walk to the place where we had the catering done and walk back. And it was because we needed to have the budget in order. Now that was the thing that saved us. And it turned out to be I think a very, very good step. I haven't heard any complaints – real complaints about it.

Actually, the managers of the venue, they were very surprised by our way of dealing with it and they said, "Oh, that was cool." They never thought about this kind of solution. Well, I'm not sure if they've actually done it since, but they said, "Well hey, we'll keep this in mind in case other people have the same complaints about money and catering and that kind of thing." And it worked out. The caterer itself, the in-house caterer was completely pissed off and mad because we had arranged to have–there was a coffee counter in the venue itself that would be open during the Saturday and Sunday but once they knew that we had figured out the catering outside. Well, the first break that we had the people outside, they knew, "Oh, wait a minute, that's their kind of thing." So they closed the whole coffee thing as a protest like, "This is not what we wanted." But it's one of those sort of kind of things that was I think kind of exciting for me about Girona, just trying to fix things and being flexible. I think that also made State of the Map kind of another type of conference. So you don't need to be in a conference where you have standard conference food. Well, you can do other things outside. So that was basically Girona.

I think the highlight now, Denver. Gosh, that was the first one outside the EU. I still kind of remember, we had these discussion whether we should actually do it because "Would there be enough people going to Denver?" Because at the time that we were having Girona that year, that was the first State of the Map-U.S. I think it was in Atlanta. I haven't been there but it was a turn up from—I think less than 20 people turned up there. So that was also like an "Ooh, gosh, so that if you have a State of the Map in the U.S. gosh only 20 show up there, and how many people from Europe are willing to step on a plane and actually go to U.S. and go to the State of the Map or an OpenStreetMap event there?" So we had those discussions whether how big a challenge it was to actually go outside of Europe. But at some point, it's like "Well, we need to go outside of Europe at some point" and Denver was a very good bid, and a very logical step at the time because after four years, being in Europe, the fifth one—it would be nice to go the U.S. And also try to get OpenStreetMap in the U.S. going.

Well, Denver is a very nice city. We were at the –I think, what was it called, the Tivoli? It was a very cool building but not always the most practical one because not everything was close nearby. But I think it turned out to be a very, very big success. Again, I think, at the time, even more people signed up than in Girona so we were getting higher and higher numbers of attendees each year. Looking back at that time, if you were in Denver, we were also looking back on a very successful kind of thing and also proving ourselves that OpenStreetMap was ready to go beyond Europe.

Say, Henk, what should we have talked about that we haven't talked about so far?

There's one subject I'd like to stay away from but it's one of those things that at some point you need to talk about. And it's more like the type of community we have, the type of people that get active and the struggle between OpenStreetMap being a very cozy project for lots of people to do some mapping and that kind of thing. So you have the mapping community on one side and you have the "apping" community or the people who actually use data on the other side. There is a kind of a tension or friction between them. At the beginning it was not really, really apparent because not that many people were actually using Open Street Map in a commercial environment. That's changing now, changing dramatically. And it's one of those kinds of things that you see lots of people have opinions about on how the commercial world should be listening to OpenStreetMap or should OpenStreetMap adapt a little bit more to the commercial world?

And that's going to be one of the biggest challenges. And you see it all through, just the kind of think like. "Is that it? Or is it just cultural differences?" I think it might just come down to cultural differences. The next hurdle that OpenStreetMap needs to face is—they started out in Europe. And Europe is a bunch of all different kinds of countries with their own languages and their own cultures and habits, but it's still Europe. It's still a group of people that work together in a special, view the world in a specific way.

Suddenly, you have the U.S. which has slightly different ideas about how the project should be running. You have different kind of problems because in the U.S.—it's not coming from little small countries where they've just basically created an old map. It has all kinds of different challenges and being a bigger country, fewer people, how are you going to deal with these kinds of things? Maybe it's a little bit more focused on actually using the data instead of mapping? So it's a little bit more, I would say, commercially-driven but there's more of a focus on 'how can we use data' instead of 'how can we create more data'. And then you have also the differences in Asia where you have lots of people that can't speak English and they have different cultures with respect to how you work things, on how you do things. And it's one of those things where if you look at the hard core community, it's centered around Europe and North America where there's lots more people outside in the world trying to do things with OpenStreetMap.

I think, it's coming back to the cultural differences that there's a lot of discussions going on in a community like OpenStreetMap. Discussions are good. It's sometimes an annoyance, but they're basically good because it keeps you sharp. And at some point, things get personal. And that's when it's not very good anymore. And we need to figure out how a community, as diverse as OpenStreetMap is, and it's for some odd reason we're still together despite all the differences and difficulties and the tremendous debates that are going on within the community, it still exists. These are growing pains; but we also need to be aware that we need be less introverted because the more we debate internal kinds of things, the less we look at the outside world and how we really would get OpenStreetMap going further.

Pavel Machalek

Bio

Pavel Machalek is a Co-Founder of Spaceknow, Inc., a VC-funded satellite imagery analytics company based in San Francisco. Previously, Machalek was Head of Remote Sensing at the Climate Corporation, which was acquired by Monsanto Co. for more than $1 billion. Throughout his career, he worked with numerous NASA observatories like Spitzer, Hubble, and Kepler Space Telescopes as Principal Investigator searching for planets in our galaxy and characterizing their atmospheres. Machalek holds a Ph.D. in Physics and Astronomy from Johns Hopkins University.

Favorite Map?

"My favorite place is Regent's park circle in London, which I think was one of the very first places I ever saw on OSM when Steve showed me the prototype."

Interview

Who you are and what you were doing before your involvement in OSM?

Before my involvement in OSM, we were students together at the University College London Physics Department. We were busy studying undergraduate physics. I was busy concentrating on my academic career and grad school while you were coming up with definitely nifty ideas for what to do with data and geo and stuff like that among many which was OSM, actually. Now, it's the biggest deal but I distinctly remember at the time that there were many different projects – image recognition, then this other department in this other building at UCL.

This is how I remember it. Who knows, right? And we were sitting at a bar right outside the hospital and you were like, "Well, dude, look, we're going to like take GPS's and cycle around and draw a map of the world." I'm like, "Huh? Why would you want to do that?" And you're like, "Well, you know, there's no open data. You can't get metadata. Ordnance Survey is expensive." And there was no Google Maps at that point. I distinctly remember there was just no Google Maps. There was like Map Quest which was really shitty. It still is.

And you were like "Look." And you opened your MacBook. I'm like, "Wow, what's that?" And you're like "It's a MacBook." I'm like, "Wow, cool." so I was like more checking out your MacBook and the circles and lines drawn in it but—yeah, and you showed me–it was like Regent's Park or something. You cycled around

Regent's Park, so it made a circle. Well, one of the parks that's like circular in the middle.

That's Regent's Park.

They made a circle. Yeah and you like made a strikethrough through it. That was basically the path through it. And you're like, "Look, it works, you know? Here." I'm like, "Okay, that's cool." So you go to one park and how you're going to get you know the whole world?" You're like, "Well, you know, we're going to build the interface and let everyone have the ability to edit it and do that." And so I'm like, "Okay, interesting." And then yeah, it was right at the end in 2003, so like June. And then I went to India for a year. And then basically we didn't talk. Well, we did talk in between but we didn't talk much until like—well, I know pretty well, 2008.

I think I hosted some photos from your Indian trip on a computer.

Yeah, you helped me post the photos because again that was before Flickr, there was no Flickr.

Before photo hosting was a thing.

Before photo hosting, you had to host your own God-damn photos. It was a pretty high tech operation at the time. Yeah, who knows? It might still be there.

And then we didn't quite talk for some time because I was miserable in grad school and you were busy getting famous. And so, I was in Baltimore, in grad school - towards the end of grad school and I'd go to my favorite anarchist café which is still there. It's called Red Emma after I think Emma Goldstein or something. She's a really famous American anarchist, like a firebrand, and would throw bombs and advocate for like women's sexual reproductive freedom or something like that. So she has a café and is run by anarchists.

And so, I go there and it's normally just like kind of hit the crowd, nothing special. But this time, it was Saturday and it's like punks in there, like Mohawks. And they're sitting around the table drinking their organic coffee or whatever. And then they have like 15 or 20 GPS units on the table, just physical units. And I'm like, "What are you guys doing? Are you blowing up Federal buildings or something?" And they're like, "No, dude. We're going at a mapping party." I'm like, "What the f*** is that?" And they're like, "Well, you know, there's this like online interface and you can like cycle around and you have your GPS on you and it makes a map." I'm like, "Huh? Interesting. What's it called?" and they're like, "Well, it's called OpenStreetMap." And I'm like, "Okay. Well, damn, you know, I heard that one before. I wonder whether it's this Steve's thing." So I look and all of a sudden I'm like, "Okay, fine. They seem to have mapped the whole world." I'm like, "Okay, who did it?" and you were not really in there anyway so I had to like Google it and like in "about it" says Steve, so I just e-mailed you It's like, "Hey, is this yours? Because I just saw some punks editing your interface." And you're like, "Yeah, it's mine." And by the time you were already in California, in CloudMade, I think. And I moved here as well, to work at NASA. And that's basically how we got re-connected. So that's essentially like the whole story in a nutshell.

I'm not a particularly active contributor to OSM. I think I did one single road in Costa Rica because it just wasn't there. It was still a jungle so I think I submitted that one. In that sense, my involvement is marginal but I might have been, if I'm not wrong, one of the first people to see an actual – one of the first polygons in there - on your laptop.

Do you use OpenStreetMap today?

We do actually. Yeah, I use it both, my personal capacity through the different apps that I use – mapping and navigation apps including Waze which might or might not be OSM but Waze does use openly available geospatial data

Waze is not using OSM.

But I use OSM for actual cities when I go travel, especially third world or like interesting places. The maps seem to be better because out here in the States or Europe everything's nice to know that this is like a water fountain and like the third sub-kind that the Germans have sub-categorized. But out in like Costa Rica, it was extremely helpful, for example. And when it wasn't there, you just added the road.

And then work-wise, professional capacity, there are some applications. For example, just today, if you want to classify imagery and find specific objects on it, you need a ground truth data set. In other words, you need a match between what's in the image and a map of that image. You can actually do that with OSM pretty well. I just read this guy's thesis who basically took massive amounts of aerial imagery and used OSM to provide the ground truth data. In other words, to tell the algorithn what is that building and what is that road. And so, that's a very interesting use case. Essentially, basically, a rasterized version of OSM with different colors corresponding to different classes of objects.

Is there anything that we haven't talked about that we should've talked about?

Well, like the surprising aspect of it. It was a huge surprise that concepts and implementations matter. The concept of this is very strong and the implementation of it was very strong to actually enable it to spread worldwide and millions of users. I guess, that's more of a question to you. It's like, what do you attribute the stickiness of OSM? In other words, the ability or the willingness of people to contribute time and effort and money basically for free. That's a question to you.

I think, you just get out of the way. You create a platform that's just good enough to do to allow you to map and then you get out of the way from anything else.

Right. And in terms of incentives? Like, if a person sitting at home watching TV, what would make him/her instead go on out mapping - for free?

Creating a sense of community so that it's not about them, it's not about the map. It's about their standing in the community and whether their map is as good as someone else's map, right? What is their relative status in this tribe?

Mm-hmm, fascinating. Going back to tribal, I love that. So yeah, because that applies to all kinds of other things than just OSM—how to motivate people to do things for free.

Mikel Maron

Bio

Mikel Maron is a programmer and geographer dedicated to community and humanitarian use of open source and open data. He has organized mapping projects in India, Palestine, Egypt, Swaziland and elsewhere with Ground Truth Initiative, and especially our flagship effort, Map Kibera, the first open source map of the slums of Nairobi. Heís a long time contributor to OpenStreetMap; and Founder and Board Member of Humanitarian OpenStreetMap Team, having helped facilitate the OSM response to the 2010 Haiti earthquake. Mikel has served as technical lead for Moabi, a collaborative data project to monitor natural resources in DRC. He co-founded the geoweb company Mapufacture (now part of ESRI), helped build the first wiki at the UN (WaterWiki at the UNDP), and generally worked on collaborative platforms and geoweb standards, with everyone from multinationals to anarchist hacker collectives.

Favorite Map?

"I'll pick kibera, Nairobi, Kenya, because I am so proud of the work of the team there."

Interview

Let's start with who you were and what you were doing before you found OpenStreetMap.

I was living in the U.K., in Brighton. I was doing a degree in Evolutionary and Adaptive Systems, which was super interesting, but I couldn't make it useful. And I had been doing open source software development for a few years after I left Yahoo. I got into maps by accident, just in the dot-com bust. I thought it was interesting and fun, so I worked on something called worldKit and did a lot of work with GeoRSS. And so, I was in those circles and was around in the U.K. That was pretty much my life at that point. I guess in 2005, I first heard about OpenStreetMap.

How did you first hear about it and what did you think?

I think I saw you and Tom Carden present at Dorkbot when you shared the poster prints, is that possible? That might have been the first time or like summer of

2005. I think it was a visualization of all the GPS traces in London. And then I kind of crossed paths with you guys in a couple of places and really, I think, understood it when Ben Russell invited us up to Nottingham and there was a—what was this workshop called? The Locative Arts space that I was getting into through Headmap and everything. The Headmap Manifesto was a big influence already, I should say that.

Do you want to talk about what Headmap was?

Everyone should read the Headmap Manifesto. It's a multi-author stream of consciousness, a visionary, evolving text which no longer officially exists on the Internet but there are archive copies you can find out there. Ben Russell, was one of the main instigators, along with Anselm Hook and a few others. Just this really outrageous vision of how technology and connections globally and new ways of looking at the world are going to upend everything. I think a lot of it was written at Burning Man or after Burning Man. It's really worth finding your way through. It's a great text that has laid, I think, an alternate vision of a lot of what has come to pass. And there's just great poetic language in it.

There's one great phrase about "searching for sadness in New York" which always stuck out for me. A lot of technology is so freakin' optimistic all the time and doesn't allow for the full range of human experience. So it's not about finding Starbucks, it's about finding sadness. And actually, over the years I've been able to make that connection, to mapping, to sadness and suffering and misery—hopefully for the better.

And I remember when we all gathered at Nottingham to basically hack together for a week. Christian Nold was there. Libby Miller and a few other folks. I remember some of us actually built something which could find sadness in New York. I think it was a visualization of Flickr geotags as map tiles and you could browse all sorts of emotion everywhere. I do believe we actually found sadness in New York that week.

So yeah, that was a pretty big week for me. I fully contemplated what you were trying to do with OpenStreetMap. I thought it was incredibly audacious, simple to understand, and it was either going to succeed spectacularly or just fail totally. I think it's the former. For many years, I wasn't sure. And we powwowed quite a bit, had some crazy plans and I think pretty soon after, I dove right in.

What did you start doing? What was your interest?

The first stuff I did was try to help out with the technical side of things. I was looking at the map visualization, tiling. The first version was basically a Ruby script rendering white streets over a Landsat background from OnEarth and it was very inefficient. It fell over all the time. And there was this crazy idea, which I guess became CloudMade, to build a business around this thing. Was it actually really like a month or two later? Ben and I were pitching a start up to the O'Reilly VC guys in San Francisco.

O'Reilly AlphaTech Ventures.

AlphaTech Ventures. And we went to Yahoo which actually turned out to be a really important early partnership for OSM; even though it wasn't exactly what we were thinking about at the time. And then I left for home, back to the U.K. And you came out to SF and ran around with Ben. I remember you also going to Yahoo but trying to call me for directions, because you guys were lost on the 101 and couldn't find Yahoo, and were super-late for your meeting with them. You also went to Google and they wanted to give you and Ben quizzes and puzzles. Yeah, that was like within a couple of months, I think, if I'm not mistaken.

And then how did we get from there to mapping Brighton?

Well, I didn't do too much mapping to start. I was mostly interested in coding here and there. When was Isle of Wight? Maybe in the summer of 2006? I biked there from Brighton.

That was huge. I mean, OpenStreetMap is admittedly a pretty insane idea. And it was an incredibly useful test to try to map an entire island in one weekend, which we did. It was also a lot of fun. And I think that's when we started calling things mapping parties. I think Ben must've—I don't know if it was Tom or Ben or somebody who came up with the name.

It was Tom Carden.

It was Tom then. I mean, this was the most – one of the more brilliant tricks of OpenStreetMap is that it made something which traditionally had been seen as a huge chore, to do surveying, to make that something that people did socially and really were excited about. Yeah, I biked there from Brighton. It was a lot of fun and an amazing way to discover a place. I got inspired from that to do some mapping myself.

Also, in the meantime, I was already involved with disaster mapping applications, mostly looking at GeoRSS as a transport format for alerts and for sharing data in crisis. I went to the European Commission Joint Research Center in December of 2005, so it's again, a couple of months later after Nottingham, and presented a bunch of stuff: worldKit and GeoRSS but also presented OpenStreetMap to them as a means for producing data rapidly after a disaster which they, I think very politely, gave very polite encouragement. But mostly just probably thought I was absolutely insane. But at least they allowed me to speak.

Can you talk about what happened at the Isle of Wight?

Well, I biked as I said and I think we rented together a holiday house somewhere on the west side of the island. There were maybe 20 people and there was one guy who lived on the Isle of Wight - David, I think, took up organizing.

David Groom.

That was the first time I had seen the "cake cut up"—Isle of Wight. And David also got the local media involved. I remember him doing some interviews, maybe you and some other folks doing some interviews with local news. We divided up the cake of Isle of Wight.

I think it was two days. I was biking around. I went out with my GPS unit. Actually, for the Isle of Wight, I had just gotten a new GPS 60CSx and just biked roads in a couple of spots, took notes on the names of streets. Other people were using different techniques. There were some folks talking to themselves onto audio recorders. There were folks driving. Some folks walking around. We did this all over the course of two days. I think the first day, we definitely ended up at a pub in the evening.

And actually, pretty much 90% of the island was covered, if not more. We didn't have a very good way of actually visualizing that except for loading all the GPS traces into JOSM, showing everyone's collective work. And that was amazing. It was really incredible to see what we could all do together over just a course of a couple of days. I think it made a lot of us, who weren't sure if it was actually possible, into believers. And it reverberated as well. I remember folks like Ed Parsons who was then at the—oh God, I'm blanking on the–

Ordnance Survey.

Yeah, at the Ordnance Survey. It was really—they even took note there. So it was a big deal and that inspired me. It inspired me to really get involved with mapping Brighton where I was living. And a couple months later, I threw a mapping party myself, connected to a conference in Brighton, and ended up getting really obsessed and biked nearly every single road in all of Greater Brighton which includes the villages next door. And finished Brighton, just at least all of the roads, the following year.

Do you want to talk about worldKit and GeoRSS?

I was really into weblogs. It really started to pickup as an interesting thing in 2001 and 2002. RSS had come out in '99 while I was working at Yahoo, on My Yahoo, and it really captured my attention as a way to easily get a view on the entire web and personalize it to your interests. My Yahoo was this personalization site that used to be a big deal and was very interesting to me, but it was only focused on content that Yahoo itself controlled. And RSS was interesting because it was potentially the entire internet. That was actually part of the reasons why I left Yahoo. They decided not to adopt RSS in My Yahoo.

It was not until 2003 or 2004 that they got on board so I just built a lot of stuff around weblogs and RSS on my own. There was Dave Winer and Scripting News and Radio UserLand. I was doing a lot of coding there and getting into Flash coding and I decided to do a little visualization of weblog updates. There was something called weblogs.com. Any time any weblog updated, it would ping this API so you'd have a stream of all the most recent posts, which weren't too many back then. And nowadays, if you checked it, it's just full of spam. There was Josh Schachter who started the Geowanking List around then as well. He did something called GeoURL, using what was first called ICBM tags in the metadata of your blog to indicate your location.

And so, you had the location of blogs. You had a relatively real-time update of blog posts. So I took both of those things and created a visualization called World as a Blog which basically just was a map. It showed the current sunlight distribution over the Earth. And anytime a blog posted anywhere in the world, it popped up in the map.

This was my first taste of internet attention. It was great. People were watching it for hours. It was one of those things that people got obsessed about. And I went to ETech that year. I got mentioned a couple of times. That was pretty exciting.

And so I decided that maps were a really awesome way to display information and also a really obvious choice. I had no background in geography or in GIS. I had had some background in Environmental Science but nothing where I really connected it to mapping. And so, I set about to make a "generic" World as a Blog, where anyone could plug in their own information. There was no Google Maps back then. There were very few ways to easily make a map on the web. So that's what worldKit was about and GeoRSS was the transport format. Just build on something everything already supports. There's a lot of software support for RSS, so just add a couple more tags to indicate location on individual posts. That's what worldKit was. It was built in Flash and I made it open source.

And I was pretty focused on building that up for a few years, for two or three years, right around the same time. I oscillated between sprints on OpenStreetMap, on worldKit and then I got into some consulting gigs with the U.N. and was doing wiki projects for the United Nations Development Program (UNDP).

It's funny that you mentioned wikis. My favorite story about wikis, at that sort of time period, was going to a meetup at a bank, like Credit Suisse or something like that in the city – in London. A guy got up and he showed some wiki software and then all the people in the audience were like, "Wow, this wiki stuff is really good." And he showed some ways that you could use it internally, like you could have a web page where people updated their own contact details instead of having to have someone in the office responsible for going in and keeping that up to date. And the very first question, I remember so vividly, was

Mikel Maron

"This wiki stuff is great but how do I make sure that I can lock it down so my employees can't edit anything?"

Yup. I still grapple with this question with OpenStreetMap. Even today, it's still, "Well, you know, how do you make sure that no one is subversive?" We're still fielding that question. Of course, I don't know when you first started describing OpenStreetMap as the Wikipedia of Maps. It's probably early on, the shorthand to describe it. But I've been saying lately, in a few years, we'll be calling Wikipedia the OpenStreetMap of encyclopedias.

That'll be wonderful. So how do we get from here to that pub meeting we had with Tom where we sketched the design?

Oh, that was early. That was, I think, a couple of weeks after Nottingham, we gathered in a pub near King's Cross, right? And that was the first thing I worked on. Yeah, Tom did that sketch and then I—

Do you want to describe that sketch?

It was on a napkin. There was a logo in the upper left hand corner. There were a couple of tabs, I think, for browse and edit and maybe GPS traces. And then there was a big map. Maybe a search box and a login link in the upper right. Basically, that was the layout that we had on OpenStreetMap until the MapBox redesign, which is not too far off from that anyway. I don't think we had a slippy map before that. I don't even remember what the interface was like before that. I don't know if there was a slippy map or anything. But that was late 2005 when we did that.

And then how did we get from there to the pub meeting with Artem where he like showed those maps in color?

I know. That was amazing. Like I said, the main rendering was Landsat images with white lines for roads. And somewhere along the way—what was this called. I remember tiles at home and I can't remember what that—

Osmarender.

Osmarender, yeah. So, this SVG transformation. There was an XSL transformation file to transform OpenStreetMap data into SVG which then could be rendered to an image. There was this elaborate architecture, basically like SETI@home, Tiles@Home to do the rendering. It was fine. It was good. I think, Artem—I don't remember how he popped up. He might have just popped up on the mailing list and we went to a pub in London. He came down from Oxford and he broke out his laptop and showed on-the-fly rendering, zooming from all of the U.K. down into Oxford—rendered on the fly and I was really astounded. It was really amazing. And he was just like, "Oh, yeah." He was so blasé about it. After showing the most amazing map rendering—the map rendering problem which we had really not cracked for several years, he just said, "yeah, sure, it's just a bunch of C++ and OpenStreetMap is good data, so here it is." And that was awesome.

Do you want to talk about the metadata that existed in the map but just wasn't being shown?

Yeah. The first version was literally just anywhere you drew a road, there would be a white line. The roads wouldn't be colored differently or rendered with different widths or different classifications. You wouldn't see the names of the roads. If there were any amenities, any points of interest, those were not rendered at all.

I think Osmarender started to do that but it was a little clumsy, I think, to work with and things like the labeling were broken up and not smooth. Artem was very obsessive about things like labeling and dealing with collisions and the maps were just beautiful. They just looked really great and it was very fast. It was fast enough that you didn't need to have this distributed architecture. You could run a couple of servers yourself in order to produce tiles.

So what happened next?

What happened next? I definitely was talking a lot about what became HOT. And – do you remember this meeting?

I remember having a meeting down in Brighton, maybe it was 2007, with you, Nick and Ed Parsons right before he left the Ordnance Survey for Google. We made steak burritos and we were plotting to start CloudMade. I don't know if the first State of the Map had happened by this point. I think it had because that's— well, I'm not sure.

I'd already done some work for Ed at Ordnance Survey and he ended up being interested in GeoRSS too. I know that there were times when you and I had some falling out, a long period when I didn't like—There were several times in the time we've known each other where we weren't talking for months if not a year. I don't know if this happened soon after this. I do remember not having regrets about not becoming part of CloudMade but also thought, "You know what? I probably should've given those guys some money." There's a whole CloudMade story there which I wasn't so involved with but I remember that. I remember State of the Map in Manchester. That was in 2007, also?

God, I don't remember, but do you want to talk about that?

I remember working with Andrew Turner on getting OpenStreetMap on the iPhone. I don't remember the sequence of this but we figured out how the Google Maps application was storing tiles on the phone. It was an SQLite database and we inserted OpenStreetMap tiles into the db. This was probably the main reason why I bought an iPhone.

It was right around that time, in late 2007. I was also the first to get OpenStreetMap onto Garmin devices. They had some funny format called Polish format or something. Anyway, that was one of the things I hacked on.

State of the Map, right? Yeah, that was great. It was in Manchester. I think I presented HOT there. That was my talk about "OpenStreetMap, a Disaster Waiting to Happen" which was also a comment on the project always being in a state of chaos, mostly technically but also in other ways.

I'd been to a lot of mapping parties by that point but just to see us come together in that format, that was the first time I really felt, "Okay, maybe this really is going to work." I think, until that point, for any number of reasons, things could've just collapsed, you know? And to have that many people come together and show all the different parts of OpenStreetMap that they worked on and cared about, not just the map but technical and social aspects. It was a huge confidence boost for the project – that first one. And then to see it happen again and again was amazing. And to see people come from so far away. I don't think it was at the first one but the second one in Limerick, there were Japanese guys who came out and, that's incredible that we're able to have such a risky attitude.

Let's talk about how HOT evolved and what happened?

That was one of my early interests and primary interest maybe with OpenStreetMap, following the first presentation of the idea at the European Commission in 2005. I mean, I was incredibly naïve, as we all were, with anything to do with OpenStreetMap but we were constantly trying to learn and improve and listen. So I spent a lot of time trying to understand more what it would really take to do disaster response and talking a lot with folks like Paul Currion of humanitarian.info who'd done a lot of work with technology in the humanitarian space and Nigel Woof of MapAction.

I was lucky that just right there in the U.K. where people who were willing to give this idea a listen and to give serious feedback. So I spent time presenting the idea. I had no opportunity to actually do it.

I went to Rome in 2007 to the Food and Agriculture Organization (FAO) which has its headquarters right in the middle of Rome. There was an open source GIS workshop there. We had a mapping party and mapped parts of Rome. It was completely unmapped at that point. I met some of the local OSM community. Schuyler was there. I started to meet people with history in the UN and field GIS. So we really got some supporters by that point.

The next big event, was when Schuyler and I went to India in February 2008, the Freemap India Tour, organized by the free and open software community in India, particularly Shekhar Krishnan. We travelled to seven cities in one month, giving OpenStreetMap workshops and holding mapping parties. I had bought a bunch of GPS units for mapping in Brighton and we brought those GPS units. We travelled to Mumbai, Bangalore, Kerala, New Delhi, Calcutta, Punjab and gave lots of presentations. Had incredibly, outrageous, silly experiences and constantly fought with the technology because this was India in the beginning of 2008.

You can't rely on power. The power was constantly going out. You certainly can't rely on WiFi, that was constantly going out. You'd have incredible pomp and circumstance everywhere you go. Foreigners were definitely a bit revered and there are also places that don't see a lot of foreigners.

And so, in Kerala, there were literally 60 people who had come to this workshop who really weren't able to understand much about what we were talking about but it was so important for them. There certainly were some people who got it, sure, and everyone did edit OpenStreetMap. Most people did edit; but it was so important just to be there. I think that was the first time Schuyler and I had to sign certificates for everyone who had attended the two-day workshop. And 60 certificates, your hand really starts to hurt when you sign that many.

And they were so happy to see us. In fact, one attendee kind of kidnapped Schuyler and I and offered to give us a tour around Kerala. Basically, we were driven by places where people he knew were hanging out and he made an excuse to stop for something, but really he was showing off that he had two Westerners in his car, which was fine. He took us to his home and his wife made us a meal, still it's like one of the most amazing meals I've ever had, home-cooked Keralan food. It was so spicy. The whole family just stood at the wall and watched Schuyler and I eat this amazing food. I'm like, "Don't you want to join us?" And he didn't let us go. He kept showing us different amazing sites all day and then wanted us to stay with him indefinitely. We were exhausted. We escaped his hospitality. It was an exhausting trip.

We had a similar reception in Punjab. Incredible time but the most insane things happened. I was very sick by this point. I got sick multiple times. They put up all these posters around about how these experts from the West were coming to teach them mapping. There was Schuyler and me on beautiful hand-drawn posters. They presented us with awards and flowers and there was local media there. This was at a technical college. we met with the entire engineering professor staff who were all very experienced Sikh professors. Very, very serious guys and we're like, "Yehaw, well, let's make a map." Schuyler and I were just silly. I was sick and so I didn't want to take antibiotics. I wanted something else—to talk to a doctor. My stomach problems became the topic of conversation among all of those professors; they have very few boundaries.

And in the evening, they said "We have a very special entertainment for you." We were brought to some place, some building. We didn't know where we were going. We got to a certain point at a hallway and they said, "Okay, go ahead." So Schuyler and I went ahead. There was a beautiful young girl student who was throwing flower petals while we walked and we entered.

We got to an inner courtyard where assembled were several hundred girls who promptly started screaming and clapping and yelling as if we were the Beatles or something. They had us sit right in front of all of them, in front of a stage and they came with snacks, tea and everything. And they then held a dancing competition in which Schuyler and I were the judges.

The students were restricted at night. They couldn't go anywhere so they just practiced performance dance with each other. Some of them very Bollywood-esque but included everything from traditional dance to the more risqué modern Bollywood, and we had to judge them and appreciate them. They had assumed that we were married which I guess we officially were at that point though not really. We were both going to be—or soon be—divorced. When they later learned that we were unmarried, they were absolutely shocked. They never would have let us watch their young women gyrate in front of us.

That's hilarious.

So that was pretty typical. It's amazing but that's what we were doing in India. The main thing to learn is well, technically, you can't rely on anything. You have to be extremely flexible and patient. This was my first exposure to see that, within open source software circles, we take so much for granted when we try to do things within unfamiliar cultures. I'm sure you are already familiar with this dynamic, trying to work with German programmers. Joking.

One of the amazing things about OpenStreetMap is how much effort has to go into not only adapting the concept so that it's understandable to a local context, but adapting how you do things completely. There are just some places where, because there's not a surplus of free time and leisure time in the same way that there is where OpenStreetMap started in Western Europe, you really have to take a different approach to growing communities. That was a big revelation there.

It can't all be revolving around pubs then, huh?

And you don't go to pubs. They always wanted to take us to go for "hot drinks", which is what some Indians call alcoholic drinks. We were always confused like, "It's fucking hot, we do not want any hot drinks." But they didn't want to drink themselves, they just wanted to watch us drink. So yea, you have to serve tea. There's different ways people come together in every place.

Soon after that was Limerick. And when I was in Limerick, I really don't know how but I got connected to JumpStart. I was going to Cairo afterwards for Wikimania, and I got connected with JumpStart International, which was mapping in Jordan. They convinced me to come to Jordan where we decided to do an OpenStreetMap project. We got kicked out of Jordan by the National Mapping Agency which is actually a branch of the Jordanian Military.

And to my horror, we went to the West Bank. We went to Palestine and started a mapping project there. It actually turned out to be a hell of a lot of easier. You could do pretty much anything there and we kicked off the mapping of all of the West Bank. That led to interest in Palestine as a whole. When there was, in early 2009, the conflict with Israel, there was a group of people who had been talking about HOT.

We decided to map Gaza but there was incomplete imagery, so we had a fund raiser to raise $5000 to buy imagery of all of Gaza from a Digital Globe reseller. This was fresh imagery. It was taken during the conflict so you could see destroyed buildings. You could see military vehicles. It was pretty crazy. We used that imagery to base map all of Gaza. That was really the first HOT activation. Around that time, I named the Humanitarian OpenStreetMap Team and started to formalize the idea.

Didn't you build the West Bank's first satellite or something?

Oh, yeah. Schuyler was also there. I was getting into kite photography, so I had a kite, a Canon camera with the Canon hardware developer's kit on it, so you could take multiple photos automatically. We went to an area outside of Bethlehem in the West Bank. It was originally a Jordanian military post when the Jordanians had control of the West Bank. And then after '67, it became an Israeli military post. And just a couple years before, that post was abandoned by the Israelis. They had left. I don't know if it was part of any of the peace negotiations.

Part of the area that was part of that post—I mean, this is very close to Bethlehem and actually close to the area that they say Jesus was supposedly actually born, he was probably born nearby—they had made into a park. A very normal park with places to picnic, ball fields, play grounds and green grass just like anywhere but it was freaking astounding to see a park in the West Bank because most of it, of course, is desert. But these everyday sorts of amenities that we take for granted are just not there in a conflict zone.

We flew a kite with camera there and we ended up flying the kite over part of the area which hadn't been re-developed, the former Jordanian and then Israel Defense Forces (IDF) camp. That was the first Palestinian spy satellite. It came crashing down. It was really windy. The camera fell off the kite, crashed, but we were able to recover the images.

How do we get from here to Map Kibera?

After doing the West Bank and getting involved with HOT, and doing India, I guess, I had built a reputation of the guy who's willing to go to some crazy place and do OpenStreetMap stuff. In early 2009, there was a BarCamp in Nairobi and some of the folks there, including Jubal Harpster, were playing around with this idea to map Kibera. Kibera is a well known slum in Nariobi, Kenya. It's famous because there are a lot of international organizations and humanitarian NGOs in Nairobi. It's a regional center.

But right in Nairobi, there are a lot of people who live in very poor conditions. Kibera, at the time, was the most well-known slum. And it was completely blank on the map. People talked about there being 1 million people living in Kibera. That's false. That's a very over estimated figure which has its own interesting story. It's closer to 200,000 to 250,000, still a lot, and it was completely blank on all maps—on OpenStreetMap, on Google Maps, on official government maps, there's nothing there.

And of course it's a slum so it's not easy to map. Especially from satellite imagery, you just see a sea of metal roofs. It's hard to see anything. It's hard to go there and it's hard to know what's actually mappable there. So I'd heard about that idea and I took it seriously. Jubal and I started working on what it would take to map it. And built up a project over the course of a few months, got it funded by JumpStart and went there along with Erica, now my wife, to Nairobi in late October 2009. And that was the start of Map Kibera.

I had some connections also with Ushahidi, which had started in 2007-2008, about the year before, through Where 2.0. Andrew Turner and I had connected a lot with Ushahidi and had been talking with them about a lot of things. They're based in Nairobi and I met some of them the year before. They were really helpful in getting things started and providing a home for what we were doing at the iHub; which didn't exist at the start but at least, over the course of the year, started to be a space where we could work. And they were helpful with connections into Kibera because I mean, I had been to Kenya once. I knew nothing about Kenya. I certainly knew nothing about slums and they were really helpful. I've tried to read as much as I possibly could but to be honest, I was completely scared about what we were trying to do and what I would find in a place like Kibera. But as it turned out, I should have been scared for entirely different reasons.

And so, what were the goals there? What did you achieve? What was the impact?

The original goal was really small. Six weeks, we were in Nairobi and probably mapping for four of those. We were just going to—it was really naïve—make a map. The original idea was actually to share it with both OpenStreetMap and Google Maps, though I thought that was useless. Various people involved thought that was important. It turned out that Google Maps didn't take the data at all.

But basically, the idea was to map Kibera using the OpenStreetMap process and then leave behind equipment and a trained group of people to start a mapping community or a mapping chapter who would then use it for what was important in Kibera.

This was a question from the start. "Well, what is it good for to have a map of a place like Kibera?" Everyone knows where everything is, those who live there. We didn't have very good answers for that question and were pretty honest about it.

We went in saying, "Well, this is a conversation we're going to have. We have a feeling that OpenStreetMap is incredibly useful in a place like this. We don't know exactly how, but together we can discuss and figure out what that is." The idea was to actually do the mapping, and then to have these kinds of discussions, and let it grow and become something amazing on its own. Basically, fly in, catalyze something, and fly out after six weeks. And that was over five years ago.

One of my memories is mapping where toilets are and use cases around them, where to put infrastructure in if you're an NGO, or am I just making that up?

You can say, "Well, it's important to know where the paths are, the basic facilities are", but it has a pretty vague impact. How does that actually translate into something which can be useful for the community itself? We were very careful about not being extractive. There's a lot of data gathering which has occurred in Kibera over the years, lots of surveying. And Kibera was still the same and the people of Kibera had no more actionable knowledge about their place than they had before.

And though OpenStreetMap is open data and accessible, you really have to go great lengths to make it accessible in a place like Kibera. We wanted to focus on making sure that whatever was collected was then shared back with people who had collected it - with people in the community. And that's really where it's the hardest to figure out because it's certainly not just having it in OpenStreetMap. It's definitely better than what had happened before, where data would be collected and then siloed off into a proprietary database of an academic or a development agency, and no one would ever see anything. You can put it online in OpenStreetMap. That's great but one thing you can do is print maps, and that's certainly very useful and that's actually really meaningful. Printing is really the most meaningful map to anyone in a place like Kibera because you actually see it physically.

But then thinking through exactly how to have impact—it's like you pull on a thread and then it just keeps coming and coming. You then encounter extensive problems with how information is managed in a place like Kibera, how people communicate with each other and organize. You get into deep issues of a place which is very marginalized and impoverished, and in some ways ignored by the government, and has a lot of NGOs which are not necessarily coordinating with each other, all by just asking the question of "How can a map be useful in a place like this?" You become aware of the myriad problems that a place like that faces which have nothing to do with OpenStreetMap or technology. It has a lot to do with education and with politics and power. So a lot of the work is trying to understand how the community functions, and the mapping itself is important and the data is important, but it's more of a vector to get to the place where the conversation that needs to happen can really happen.

I remember after the first month or so, we quickly decided we needed to come back and do more. We got funding from UNICEF to do that. Part of that was more focused exercises with particular groups. So, we had water and sanitation, education, health, and security as four themes that we worked on; and we collected data particularly on these themes in OpenStreetMap and then printed physical paper maps and held map-drawing exercises, tracing ideas and experiences on top of them.

For water and sanitation, we gathered together all of the local NGOs that were working on toilets and on water. This is a place that has no access to formal water or sanitation services. You have open sewers; you have pit toilets. To get water, you have to buy water from a vendor at extremely inflated rates. It's a mess. It's probably the core problem of the place besides the lack of political power.

We gathered some of these groups together and did this exercise where they map out the issues that they're facing. "Where are you doing interventions?" So, everyone's mapping out where they were putting in toilets or water services. And all of these groups, they've been working on the same objectives. They knew each other. They had been in meetings together organized as networks or coordinating bodies for all of the water NGOs in Kibera, but they never had a substantial, direct conversation with each other about what they were doing, and how could they work together better. They just had never been in that kind of setting. Kenya's a very formal country. Political culture is very formal. It's like Britain from the 1950s and—

Well, that's basically when Britain left, right?

Yes. Things of course have changed, but there is still a lot of ceremony and these NGOs just never had time to actually talk to each other and come together. After 10 minutes, the conversations just took off. It was really great to see how presenting this whole view of the place - a place that had never been mapped before, a place that everyone talked about and worked in but had never really seen as a whole, seeing that together was a conceptual shift for understanding of that place, and a kind of pride in that place.

I mean, imagine growing up in London and never seeing a map of London. What would you understand of it? You'd have a certain understanding but it would be very hard to have a global overall view, to try to adapt on a different level. And so that was very exciting to see.

I think that ultimately is some of the real impact. I mean, of course, there are real improvements that have been made to water and sanitation work based on the map, siting of different interventions but it took a few years until we could say, "Yes, this map has actually made an improvement in Kibera."

It wasn't until the 2013 presidential elections where Map Kibera mapped all of the polling places, mapped all of the election districts and produced paper maps that were distributed to everyone who was responsible for the election security. No one had a map of where the polling places were and everyone was very concerned that violence would erupt again, like it had in 2007 and 2008. So simply distributing that map actually made a real impact.

But Map Kibera also had and continues to have still like a whole social media and video and journalism side. There is Voice of Kibera and Kibera News Network. Voice of Kibera is based on Ushahidi. Kibera News Network is a YouTube channel. These build off of the maps, and give voice to the concerns and issues of the community in a narrative form. So they organized for the election. They did interviews with all the candidates for the parliamentary seat. They held debates and public forums. This is the kind of basic stuff that happens all the time in democracies but it was new for Kenya. Kenya, especially in a slum, did not have this sort of political culture.

And so, that was great to see, as well as specific examples of facilitating security incident alerts to the proper authorities. Things did not blow up in the same way as they did before and it was a largely peaceful election. Map Kibera helped raise awareness in a different way, which I don't think we had seen before. It wasn't until about three and half years into the project where I could say, "Here's the result. A real solid result." That was the first time.

Can you tell the story about trying to teach people in India how they're mapping and they all mapped the same thing.

I think that did happen in Kerala, in India. I just gave them an example of a place they could map in the area. They all took the exact same route that I had described. They just didn't realize that it was a real thing they were doing. They were used to doing things which were only exercises for learning. They were still students, so they didn't actually have to deliver anything so that's probably the confusion, where they all took the same route.

It's funny because, actually, I remember at Microsoft, there was a team in India of like hundreds of people. And literally, all they did was trying to find out what was going on at headquarters and trying to copy it and go and re-implement.

I think a lot of the way things are done in Indian schools is rote education and repetition. Kenya's the same. The basics of OpenStreetMap require self-motivation and, for lack of a better word, entrepreneurship, which is not very well-ingrained in education in places like India or in Kenya.

And I think in places like Palestine, it was a lot easier for mapping to happen because they were a little more educated. People in Palestine are very well-educated and they're very entrepreneurial by necessity.

So, it's not only that you're trying to introduce this new way of doing things but you're almost trying to undo the entire way they have learned their entire lives. "Just do it. It's okay to make mistakes. It's real and you have to learn it for yourself." That's one of the intangible things you have to deal with.

Do you want to talk about Katrina and the bridge of course, and then differences between Katrina and Haiti?

There's a story from Katrina where Jesse Robbins was there as a volunteer and had set up camp near to a bridge that had been totally destroyed by Katrina. And he noticed that lots of folks were driving down the highway towards the bridge, then getting out of their cars, very confused and uncertain about what to do. And he would ask them, "Why are you coming down here? The bridge is out." It turned out that the Red Cross had sent them that way, had given them directions to evacuate New Orleans via that bridge. So he called the Red Cross and reached one of the volunteer operators there. And he told them, "The bridge is out. Please don't send people down here." And the operator said, "Well, are you sure? Because I'm looking at Google Maps right now and we see a bridge there." He pleads, "No. There's no bridge here. Don't trust the map."

So that one person was convinced but there was no way to communicate to all the other operators and volunteers. People at the Red Cross had no effective way to know that the bridge was out. And so they continued to use Google Maps and they continued to send people down that way. I mean, the thing about maps is that they're always wrong. The map is always going to be behind reality.

A lot of people see Google's name on it, or some other authority's name on it, and assume that is the truth. The amazing thing about OpenStreetMap is that it exposes that data is often wrong and the only way to make it better is improve it yourself. The best thing that could've happened in New Orleans was that they used OpenStreetMap and just edited the bridge and that would have been the end of it. It actually was present on the map for several years. Eventually, they did take the bridge out of the map. It took about a year and a half. But about six months later, they had reconstructed the bridge in one direction. And so, shortly thereafter, and this was when Navteq and TeleAtlas were the primary data sources, it was out of date again.

And the process, from the time something changes on the ground to the time when you would see it in the map, the chain of things which would have to happen, it could be over a year or longer, or even never. And now, something that happens on the ground could be reflected on the map—in OpenStreetMap—in minutes.

Several years later, with Haiti, there was a different circumstance. Everyone knew there was no good map data available for Haiti. There was nothing that anyone could use, and because Haiti is so close to the U.S., there was a huge response directly from many U.S. institutions, NGOs and the government. No one had any maps of Haiti but you had the OpenStreetMap Community. HOT responded very quickly and got access to imagery very quickly through Crisis Mappers Network and Digital Globe and Google, and everyone opened up imagery which allowed us to actually map what had changed.

There was no other option and it was an extremely urgent situation. When you're in the middle of that kind of situation, you don't have time to think or question too much, you just have to act. And so, if you are a GIS officer, an IM officer for MapAction or for one of the U.N. entities which is responding, you don't question, "Well, can you trust OpenStreetMap because anyone can edit it?" Rather it's "Yeah, that looks good enough. We're going to go for it." And so, it's a very different circumstance by that point where there was really no other option.

I wonder, if Google Maps had had an okay map of Port-au-Prince, well, probably everyone would have just used that, I don't know. Or if the National Mapping Agency of Haiti hadn't been destroyed, things would've been different. Because even the meager information that they had would have had that stamp of authority. Even if OpenStreetMap was better, responders probably would have reached for available "authoritative" data. And so, it was a tragedy of circumstances which allowed for a new way of doing things to sneak in.

Jumping back to the bridge, it's actually quite funny that part of that bridge story is actually in economics textbooks because next to it was a CSX railway bridge and the CSX bridge was back up within a month but the government bridge took an awful long time to get back.

Yeah, true. I don't know if you could entirely attribute that to government. It's probably a bit more work to open an automobile bridge. The railway bridge maybe was not as—but anyway, yeah. I'm glad it's famous for even more boondoggles than the Red Cross one.

Do you want to talk about Andrew Turner and companies and what ended up in FortiusOne?

I met Andrew through the Where 2.0 and FOSS4G worlds. Except for Chris Schmidt, he was probably the most energetic person I'd ever met. He still talks pretty fast, yet he talks significantly slower now. He was one of those people whose thoughts were—he actually tried to allow his mouth to keep up with the speed of his thoughts and so there would be humongous jumps in his thinking. He's an impressive guy. A lot of fun.

We started working on this idea to build a GeoRSS aggregrator so you'd have a place which would bring all of the geographic knowledge on the Internet together in one place, in order to do interesting things with it. That was called Mapufacture. We worked on Mapufacture as a startup for about a year and a half and did a mix of consulting and then started getting into talks with VCs.

I decided to go a different way. I decided to focus more on humanitarian work and international development. I was not game for the commitment that a startup really involves. I think around then that FortiusOne acquired Mapufacture, which was cool. And Andrew just started working with Sean Gorman and them. Lo and behold, a few years later, FortiusOne which was building some really good stuff was acquired by Esri. And so, Andrew's now working with Esri and leading their good work on open data. It's been a long journey.

What about Where and then Where Camp?

Well, O'Reilly Media was organizing very cool conferences in the tech world like emerging technologies or ETech which is where I first showed World as a Blog, and Web 2.0. And so, all the really cool things were sort of gathering around these conferences. And then Nat Torkington, I think, started Where 2.0 because they recognized there were a lot of things happening in the geo-space. I think Google Maps had come out, and all the excitement around the Google Maps API and OpenStreetMap of course and GeoRSS. Mapping as an industry has been around for a long time and GIS was advanced but it was very different, disconnected from what had been happening on the web and not fully embracing or understanding what the internet was about. And so, that was the niche of Where 2.0.

The first one was pretty amazing. I presented there and met a lot of people. Maybe the second year or so, we started holding WhereCamp on the side of the conference. It became a lot more interesting and a lot more fun at the WhereCamps. The first one was at Yahoo and then at Google and we did some really amazing hacking. One year, at Where 2.0, Google Streetview had been launched. And at the WhereCamp, the next day someone had, similar to how Google Maps was reverse engineered and then cracked open, someone did a similar thing with Google Streetview and we had an amazing session of people who were doing crazy shit with Google Streetview images all from a yahoo.com IP address. We were all at the Where Camp at Yahoo. So suddenly, within days, Google is going to see a huge spike of basically hacking coming from Yahoo. It was all just kind of fun but we just did all sorts of experiments and connected with loony people. In later years, it got very practical.

I was just looking at some of the history of HOT and in 2010, this was a few months after Haiti, there was a Where 2.0 and WhereCamp where, of course, Haiti was presented and I think I presented Map Kibera. We had a meeting of some of the key people who would become HOT as an organization. We sketched out the requirements for the tasking manager, which is now the HOT tool for organizing activations and responses at scale. So that was actually, just with a bunch of people around the table like Kate and Schuyler and others - actually invented the tasking manager at a WhereCamp.

Do you want to talk about GeoWanking at all?

Geowanking is a mailing list which used to have a lot of traffic on it and purposely had a name that you couldn't say to your boss. So it, by its very nature, was going to stay a bunch of hackers. That's where I first, I think, connected with Ben Russell and heard about HeadMap and all of this stuff was through Geowanking List.

What are you doing today?

Today, I'm a Presidential Innovation Fellow (PIF). I serve at the State Department with the Humanitarian Information Unit, working on MapGive, which is a key institutional supporter of the work of HOT and of OSM. The pretty central thing that they have done for a few years is provide imagery services based on the Digital Globe catalogue, for humanitarian response. So if there's an emergency and there's a need for imagery, HOT can get new imagery for tracing. And MapGive has expanded into general promotion of OpenStreetMap within other government agencies. You have groups like USAID who are using OpenStreetMap in their programs, the Peace Corps, the World Bank which is also here in D.C. and others. OpenStreetMap is a central part of what they do. So that's part of the job, of being a PIF at the State Department, as sort of being maybe the "Ambassador" to OpenStreetMap. To help big organizations and institutions, which function in very different ways from the OpenStreetMap community, understand how OpenStreetMap is much more than a data source. It is this group of people who have a way of doing things which goes all the way back to the very early days at Nottingham and at the Isle of Wight and everything. So there's a lot of translation of culture, to figure out how institutions like those can be better citizens in OpenStreetMap. That's what I'm doing today. I never imagined that I'd end up here.

Have you met Obama yet?

Not yet. I've seen his office but no, it's not guaranteed PIFs will meet the President. If we manage to do something awesome enough, that's important enough to the President's priorities then maybe we will see but—

You should do some of your drones, clearly.

That was not my drone. It was just a coincidence that the night that a drone crashed at the White House that I had taken a tour of the West Wing earlier in the day.

What should we have talked about that we haven't talked about?

Mikel Maron

I think it's amazing that you're doing this. I'm really excited to see all of the stories of OpenStreetMap together. Here's one thing that I still don't have a real good understanding of. How have we managed—you probably have an idea—how OpenStreetMap has managed to stay together and ultimately to win. It's obviously a good idea but it wasn't a unique idea, you know? Like, there were other people who had a similar idea—but that wasn't the important thing. The important thing wasn't necessarily the idea— it was a very good idea to have but there have been a lot of attempts also to copy the idea.

I'm still trying to figure out what is the unique set of things about OpenStreetMap that have allowed it to work and to flourish. One thing that I often come to—and someone, I guess, will be interviewing you which is good-- is your role in the project, in the way you approached it, and you have created the space and the focus for it to happen.

There's probably a year when you and I didn't talk because I was so pissed off about something that happened. I think it had something to do with the scholarship program and how CloudMade was managing the money. It was frustrating. Later, I got into the role of being in a position of leadership with Map Kibera and with HOT, of which I'm not the only leader but I definitely feel a greater weight of responsibility to those things. Suddenly, I realized the pressure of that situation, the intense difficulty of balancing so many people together to do one thing, especially with some of the basic ideas of OpenStreetMap are that you can do whatever you want, to some degree.

And so after I'd gone through some of those experiences myself, in the really difficult position of leadership, whatever that means, I understood more about your position. And it doesn't mean that you're the only leader but you are responsible. There's f**king up, and letting people down, and not knowing everything and getting blamed. That's a hard position to be in. So I don't know if there's anything more to talk about that right now but I certainly hope that whoever interviews you will dig into the challenge of leadership. One of the first things I remember you saying in Nottingham, you were ready to give up, or at least that's what you expressed to me. Maybe, it was just an idle thought. No, seriously, you were like, you weren't ready to give up but you were frustrated—OpenStreetMap was already a lot to take on, and this was when we had just met, in 2005.

And of course that's the kind of sentiment that appeals to me. I was like, "Oh, well, I can help you." So maybe you were just trying to get me involved...I hope that your interview touches on some of that stuff because I know how hard it is.

You made me think of a few things. I think, the very beginning, the main difference with OpenStreetMap and everything else was universality. That it was things like - it was like freemap.il where you could map whatever you want as long as it was roads in Israel. Or it

was Nick Whitelegg's next project where you could map whatever you want as long as it was footpaths in England. And OSM just got out of the way and just said, "Look, why don't you map anything anywhere?" I think that was a major thing.

And then also the way people thought about it as a technology project versus a people project. Let's use WMS or some big complicated protocol rather than just very simple APIs and focus on letting people map instead of GIS professionals map.

I think being very happy to make failures and not care whereas most people in the world just are not very good at failing. And then also, willingness to evolve. So, if you look at most of those other projects, they never evolved past version 0.1. Whereas OSM went through, while I was doing it, three major re-write from Java to Ruby to Ruby on Rails. What else? Being an asshole helps.

That's true. Most people think they have it all figured out from the start and aren't willing to change course. And when it doesn't work out they are unwilling to persist. Certainly, being willing to persist through all of the hard times is valuable. And being an asshole. That is one way to put it, but another way is the willingness to make the decision and accept the consequences.

We didn't talk about the board or scholarships. I don't know if there's anything you want to say?

Oh, scholarships, I don't think there's much to say. It was awesome to bring everybody together and I think some of the best talks at State of the Maps were from scholars. It was great seeing people's minds blown with how OpenStreetMap happens in different places, and that was due to the scholarships.

The Board –well, this is another place where none of us had ever really worked before. Certainly, I'd never been on a board or started an organization or had any idea of what it takes for something like that to function. I don't think I understood that until actually going through the process of HOT and Map Kibera. That took a few years to understand how those things work.

The only thing I'll say, the thing about being a technologist, the great thing is that you think you can do everything. But there are other ways of doing things that are not natural to open source software people, and it takes failing many many times in order to get it right.

Certainly, with OpenStreetMap, generally, there is still a lot to get right. It'll never be perfect, with the Board and with all of the organizations around OpenStreetMap. We only saw the first official local chapter yesterday. Things that can happen fast, like mapping, work well under a certain sort of structure. Things like legal agreements don't function very well in the open source structure and trying to fit the two together is really hard. It takes a lot of time and effort. And there's not a lot of good examples either to draw from.

Randy Meech

Bio

Randy is currently CEO of Mapzen, working on open source mapping tools and services, where he focuses a lot on OpenStreetMap. He was formerly head of engineering at MapQuest, where he launched the OSM initiatives with the great team there. Randy has moved between startups and larger companies through his career, spending five years at Google along the way.

Favorite Map?

> *I've always loved the original map styles Stamen did for CloudMade, especially "Midnight Commander" http://stamen.com/clients/cloudmade I'm happy to be working with some of the people who made these today!*

Interview

Why don't we get started by asking who you are and what you were doing before you knew OpenStreetMap.

I've gone between big companies and small companies basically during my career. I like swinging between the two. I was at Google for about five years, not working on maps but working on ads. I started my career in tech at the very end of the 90s, during the dot-com boom, so I was very, very fortunate to go to Google when I did because they were the only place hiring at the time, which was great.

And I was there when they launched maps but I didn't work on maps. I did see the early versions of the maps API, but I wasn't super interested in mapping at the time, although I definitely used it in some of the internal tools that I built. Then leaving there, I did a couple of startups that fell apart that no one would have heard of. Eventually, I went on to be the first engineer at Patch which was a hyper-local news and information service.

We were trying to figure out a way to replace the newspapers that were falling apart in local communities, and to do that online, and the plan was to hire a full time journalist in every community. Pretty early on, one of the problems - since we were using Google Maps when we were there - this is around 2008 and we were testing this mostly in suburbs and communities around New York City. The editors were complaining that there would be a missing park or a missing church, some really important piece of data that defined the community. As you loaded the homepage, it was almost like the map was part of the logo of the community. It was what made the page different for every town, so it was important.

Of course, fixing that in Google, you have to email Google. They would take a year to fix that kind of thing. That's when we found OpenStreetMap, me and Brett Camper who's actually co-founder at Mapzen now. He was the head of product and I was the head of engineering, and we built our own map deck, got to know you, Dane Springmeyer, and a bunch of the people who are now active in OpenStreetMap and related technologies.

We set it up. We worked with Stamen Design, with Mike and Eric and everybody. We were able to design cartography that fit our brand, which was really great. And then we were able to solve problems that the editors had. You could edit the map, and I think that's what we did, and then we solved problems that the editors had.

I'm not sure how important that was to the business, but the key thing was that I got hooked on it, Brett Camper got hooked on it, a bunch of people wound up getting hooked on OpenStreetMap. Maybe it was a diversion. I think it was kind of a cool product feature but the more important thing for me is that we sort of discovered that. And then we were acquired by AOL. I was just really interested in working with MapQuest which had 50 million unique visitors per month on its website at the time and was, certainly, the biggest thing that I had ever managed. Not including Google, obviously, but I didn't really work on the public facing side of Google. In terms of consumer sites, it was very large and I was really interested in doing something with that. I was just really interested in doing something with open data and services.

So the month I went in, ironically was the month that Google Maps surpassed MapQuest with traffic. So it's kind of ironic, having come from Google, like now I'm there at MapQuest. Strategy wise, I was becoming the head of technology. We still had a general manager there who was worried about the business side and I didn't really have a solution for the business side. It's seemed like it would be really good idea to see what support we could give the OSM community we'd seen at the early WhereCamp and at State of the Map and things like that, and see if there's some way to combine the resources of a larger company with just the raw energy and interesting nature of this up-and-coming, open-source mapping community.

What about the big MapQuest fund for OSM?

It was a million-dollar fund to benefit OpenStreetMap. I was definitely naïve about how big companies worked and how OpenStreetMap worked. I was naïve about a lot of different things. So basically, we announced that. I think in the end, MapQuest put more resources behind it than that. At the time, I didn't really know if there was a body that you could give that money to and create direct improvements. So I didn't really know what the mechanism was going to be there, how to actually deploy that money.

But we certainly hired a lot of people to work on OpenStreetMap. We contracted people like Artem from Mapnik who worked there and did a lot of improvements to Mapnik, hired Brian Quinion from Nominatim, well I think he's actually still there, Matt Amos who's still there, and a number of other people. So I think a lot of that money wound up going into hosting the services, like they still actually host a free Nominatim service that I think is pretty popular. The tiles, we did some open aerial tiles. I know it's still around but I don't know if it's strategically a big focus.

I was actually at breakfast with someone this morning and they brought it up as like key moment in the history of the staff as proving out how a big company and the staff could work together. I'm proud that all that happened

What was your reaction the very first time you saw OpenStreetMap?

It was, "Oh, this is fantastic." I remember it was a blog post. I can't remember who wrote the blog post. But there was a blog post saying, "So you want to create a map, and here's what you need." It kind of broke it down like it was an engineering focus, like you need - first, you need a data source and data sources are really expensive but here's OpenStreetMap. And then we started looking at it and saw that it was actually viable. It was amazing. I guess it had been around for four years before I even was aware of it. Yeah, it was amazing. It was amazing that it existed and it just seemed like this huge asset and was like, it was really exciting. The rest of the post was about how to style your maps and how to generate tiles and all the stuff you needed to do. Yeah, it was very exciting.

I guess it was pretty easy to start using OpenStreetMap in Patch?

Yes, certainly. I don't know how big the team was at the time. It was certainly over 10 engineers, and we were able to take some single person's time, a back-end person or a front person and get pretty far in terms of building a stack, doing all the stuff, working with OpenLayers at the time. So yeah, it was pretty easy to do given that right then, we were interested in it. We were able to do a lot pretty quickly.

Was that different at MapQuest AOL?

Not so much. MapQuest is pretty big. It was over a hundred engineers in different locations globally. I didn't really know what to expect, and the first people I met were in the management level. But as we got in to talk to the engineers on the ground at MapQuest, a lot of them were really excited and thought that they should have been doing this for a while. A lot of people that we met on the team, would tell me, "We're so excited we're doing this. We've really wanted to do this for a while."

When you're in a big organization like that, there are lots of places where things can slow down, whether it's on the top levels or the engineering levels. But because of so much support at the grassroots, things actually went a lot faster than I thought. I remember we were trying to launch our open websites for State of the Map in Girona, which I think was - I can't remember if that was 2009 or just 2010.

We had a really aggressive deadline for launch. I couldn't believe how quickly things went and how motivated people were to actually get the thing to launch, which at the time, I mean using the stuff today, it's still a little wily. But back then, it was very difficult to get the software set up professionally in a way that AOL would be comfortable actually deploying it, and we were able to do it really quickly because the team was super motivated.

On the management level, MapQuest had a lot of traffic. I don't know how much traffic it has now but it was an important property for AOL just given the traffic that it had. And people were really open to any idea that would help, and no one really had an idea how to help it. MapQuest traditionally made money, a lot of money selling store locators. Google came up and when I say, "disrupted it" I mean it in the most literal way. I mean that's the textbook definition of disruption. Google was totally disruptive to MapQuest.

People were just hungry for a new idea, so there were really not a lot of struggles to go forward with OSM. Even on the legal side, where I would be the most worried, they were fine with CC BY-SA. They thought ODbL would be better but there was no real problem with either of them.

There was really not a lot of friction there. The friction came from the proprietary data vendors who weren't very happy with the announcement but that came later on.

One of the large mapping data suppliers invited us to participate in a fairly rigorous audit. They weren't very happy. We were audited. At the time, the data providers were also figuring out their mobile strategy. I think it was after the Nokia acquisition, so it was unclear whether a company like MapQuest was a partner or a competitor.

So it was very natural for a company like MapQuest to start looking into alternative data providers including open source ones, but at the same time, the data providers that currently existed were also competitive with their partners. No one really knew where things were going and everyone was concerned and confused about it. Yeah, it was an interesting time.

What are the mechanics of an audit? Is an audit something that's normal?

It was news to me, and I had certainly never gone through one before. Basically, people come in on the technical side and look at your system and just make sure that you're not mixing data inappropriately or anything like that. So basically, it requires you to take a couple of engineers who were busy on other projects and dedicate them to something like this. I guess it was part of the contract which I wasn't familiar with at the time but I guess it is something that they can do.

What were the public reactions to the use of OpenStreepMap?

So I remember after launching it, we had a few blog posts from tech blogs. When you're coming from the startup environment, you really want a TechCrunch article. These are articles are currency for you. I don't know how much they really affect your business in the long run. But for you and for the team, they're very motivational.

Once we launched that, we had a number of these articles that were pretty complimentary. I thought that was pretty exciting. I think that helped in terms of hiring, so we were able to hire people who were excited about the work, which is great and definitely one of the things that we wanted to do.

In terms of the actual apps themselves, I don't think we got a lot of usage immediately except from people who knew about it. In fact, we sort of segregated them so you have to go to open.mapquest.co.uk for example to use it. Mostly because then, as now, we were worried about search quality. We were really lucky and fortunate to find Nominatim and be able to stand it up and have something there. Basically, the site was a duplicate of the commercial website that we had just redesigned to be a one-page, contemporary web map design. So we did the same exact thing and put that up under a subdomain.

People used it but mainly, we got some good press mentions from the tech press and we had more interested people to work there, which was important.

Was there any mobile use at MapQuest and OpenStreetMap?

At the time, no. Since then, yes. In fact, I believe that if you use MapQuest internationally now on any mobile app, you will be using OpenStreetMap.

When I left, we were using NavTeq. They switched to TomTom later on. I believe that it's OpenStreetMap if you're not in the US and TomTom if you're in the US at this point on any mobile device.

What was the first conference you went to with the OpenStreetMap?

I think the first one was a WhereCamp. I can't remember which one it was or if it was the one you spoke at; you probably spoke multiple times. But I remember afterwards that the WhereCamp was really eye opening in terms of just how much work was being done. It was really my first experience with a really engaged open community.

Can you describe what WhereCamp is?

It was an unconference, or at least part of it was an unconference. It followed Where 2.0, which I think was an O'Reilly conference, so very produced, very professional. It's a conference you directly recognize, and then some subset of the people from Where 2.0 and others go to WhereCamp afterwards, which was either free or very cheap. I think the one I went to was at the far reaches of the Google campus. There was an unconference vibe to it, a lot of grassroots ideas about what the talks were and the tutorials and various things that people were presenting.

There were people flying kites and taking pictures. It had this very interesting open geo culture vibe going on that I hadn't really been around before. This was before, or it was maybe after I worked at AOL, but before I worked with MapQuest. I got the idea that it would be amazing to be able to take this energy and somehow combine it with what this larger company was doing to extend it if we could.

It was a really exciting thing. I don't remember a lot of the talks. I remember Aubrey from Patch talked about what we had built on your map stack. But I think his talk coincided with someone talking about flying kites and taking imagery; it was a beautiful day, so you didn't have a lot of attendance but it was cool, it was a cool conference from what I can remember.

Did you ever get into the position of being the corporate guy at WhereCamp in your community and then at MapQuest you were the crazy open guy?

I felt like I tried to be very sensitive to the vibe of the community. I don't feel like I had a lot of the "corporate guy" problems that other people would had have when interacting with the community because I don't think anything that I've been involved with has ever been obnoxious or anything. I tried to be supportive.

I don't know. I haven't really thought this through a lot so it's probably going to be difficult to explain. I never felt anything from the OpenStreetMap community about being the big corporate person. There is definitely like a strain of suspicion about corporations in general but I just don't think that anything that I've ever done at any company has been - usually it's a concern about taking all the data and monetizing it somehow and not giving back. I think if anything, we've always like led with let's figure out a way to give back somehow - we have these open services. Let's openly contribute back to them in a way that makes sense.

I don't really feel like I ever had that that problem. Then, within AOL there was just no one who really knew what to do with MapQuest and people were aware that there was a problem having undergone a full disruption. There was this interest in open. It's going fully open, so it was disruptive to a proprietary business model. Maybe that's what you need to do in order to save this thing. Maybe you're disrupting the disruptor in some way.

For a year, things worked out pretty well with that. I mean I don't know. AOL has its own issues that I probably don't want to get into here. So for a year, no one really had any other ideas. This was at least a theory, so I didn't really have a lot of problems internally. Again like I said, the rank and file engineering was very excited about this and very motivated about it and something that a lot of people thought we should have done it for a while. So yeah, there wasn't really a lot of controversy either place.

How did that meeting go? Did you just sort of get a bunch of people together and say, "Hey, I really need a million dollars to—"

Well, it was me, Christian who was the GM of MapQuest, with whom I'm still in touch, and Stella Lorenzo who now actually is at Google. We were talking about what to do and I was pitching this as a concept. Tim Armstrong, the CEO, came in and he said, "Hey what's going on?" I said, "Well we are thinking about backing open data and open source at MapQuest and seeing how it goes." He said "Okay. What do you need, like a million dollars?" We said, "Yeah," "And so okay, done."

It was very simple like that and it was very Tim. What's great about Tim Armstrong, he just cuts right through it. Get a number that's impressive and then leave the room. Then we were sitting there like okay, now what? Now what does this actually mean? Are we writing a check for a million dollars? Are we doing standup services and hiring? That's what it ended up being in the end.

At the time I didn't know what the organization behind OpenStreetMap was. Whatever I thought, I certainly was wrong. I was certainly wrong and continue to be wrong to this day. I wasn't even sure if it was something that you could write a check to to have data improved or help in some way. Of course you can do that and we did do that a number of times to help with servers and things like that. I didn't really fully understand how the community worked. I didn't fully understand - I was totally ignorant about like any issues about companies which I think was probably a benefit. Yeah, so I didn't know. Yeah, that's basically how that meeting happened.

So then how did you end up leaving AOL?

Let's see. Yeah, I stayed for about a year after MapQuest. I mean the thing about AOL, it was in a turnaround and I feel like everyone who went there, every single person, at least in the management level, was new at this.

Tim had come from Google. It was very much experiment to try to turnaround this large organization with tremendous assets that was having problems. This was certainly new for me. I had a thesis, "Hey let's play with open data with MapQuest." I don't know if that was enough. Business wise, like when you have been making a ton of money selling store locators and that business evaporates. Whether or not you're doing stuff with open data. I don't know if it'll ever really fix the missing revenue stream that has vanished.

It was very tricky to try to turn that business around. Later on, they were starting to focus on the Huffington Post acquisition which had recently happened. I just wound up leaving. Basically priorities shifted. The priorities were more toward publishing content which I just personally wasn't that interested in so I wound up--

That's like Endgadget and TechCrunch?

Huffington Post in general. I can't remember when Tech - TechCrunch came in a little bit earlier. I end up leaving after about a year because, I really liked the idea of working with the staff on the tech side and just having the freedom to do that.

It's kind of a big regret for me to be honest, I really wish that I had stayed longer. In terms of my career, I stayed five years at Google. Other than that, I have stayed two years at places. That tends to be what I do. I'm really aware now that you really need to be at a place for much longer if you really want to have any kind of impact, any kind of sustainable impact, at least through a company.

Of course open source projects are different, they're not a company. You can stay involved for longer. I really regret having left MapQuest after a year. I don't know if it would have worked out had I stayed anyway. I really wished that I had stayed and tried to work things out, and kept focusing on this stuff. Whenever I talk to someone who had been at MapQuest from the time, they always say, things were really great when you were there. We were really focused on what I thought were the right things. After you left, priorities shifted and like, people weren't necessarily so focused on that stuff.

I think people were kind of upset about it. I feel like if I had been able to stay there for a few years, we could have improved things a lot both on the data side and on the software side and on the product side just in terms of professional products using the data. I sort of regret that in the end. At the time, it made a lot of sense.

So talking about publishing, just getting back to Patch. I don't know whether this is true but at least the mythology is that Craiglist killed the local newspapers by making the classifieds both a lot cheaper in general and also a lot more available and more up to date because it's on the internet instead of a daily paper or weekly paper. If that mythology is true, did you ever think about, instead of starting at the journalism, starting with Craiglist and sort of bolting on journalism, do you see what I mean?

Yes, we talked a lot about classified. It's funny, this was the first time I started to think about this problem, and since then I've seen a lot of business models on the startups coming up trying to tackle the same problem. Often times, they start with looking at Craiglist and saying, "Isn't it ugly, can't we do better than this? Can't we structure this data better?" And many, many businesses get launched planning to do something like that. Craiglist actually works really well, it turns out, and it has totally reshaped an industry. Whenever I see businesses like that coming up now, it's always - I never liked to give negative feedback on things, but well, maybe that's a little harder than you think.

We always had some kind of classified strategy. Coming from Google there was this sort of self-served ad type, an AdWords like thing, a product that we had built pretty early on that we didn't really go far with. The thing about Patch that I liked, which is kind of funny because Patch had several large scale, very embarrassing failures later in its life. In the early days, the reason that I was attracted to it and the early team was attracted to it was the concept of hiring a local editor. You could actually make money. Maybe it's not an internet-scale business. The economics of it are such that you could hire a full time editor, sell ads in the community, and at least make some kind of profit, then as you scale up this network nationally or internationally, suddenly national brand advertisers are interested in that.

Something that you can make a little bit of profit on locally, you can then scale nationally to say nothing of what you can do with classifieds if that happens to work out as well. That was the business direction. But there was also this social benefit of replacing these newspapers that are vanishing, right? There are police departments where no one is writing about what they're doing. One of the early Patch articles that I thought was really cool was a police department that spent a bunch of money buying new cars when maybe they didn't need to buy new cars. They published an article about this and this suburban police department was just totally not used to that kind of scrutiny. It just seemed really cool that we're doing this thing that felt like a really good use of the time and resources if the thing could become sustainable business.

Sunlight is a very good disinfectant.

Right. That wasn't what every story was about. Normally, they were about this business opens or the cat is stuck in a tree and the fire department comes. Once in a while the fact that Patch was there was really interesting. There was a school shooting in New York where one of the shooters was from New Jersey and something else had happened in a third community, and Patch, by virtue of having an outlet in each one of these communities, was able to do really interesting reporting.

Again, it shows the value of it at times when it's needed. Yeah, it was cool. It was difficult because, scaling so fast was part of the model. I don't think it had been proven out to scale that fast. It would have been nice to stay in a few communities and figured out a way to make classifieds work before scaling so fast. That's in hindsight. That's really clear.

How important was the map to the Patch?

It was important to the editors who were complaining about the missing parts, because for them the map was the logo. It was like the identity for their community, because the map was so prominent. Then it became really important to the engineering and product teams who were very interested in OpenStreetMap and motivated to do that.

They didn't keep the map up for very long after I left, so I don't know how important it was in the longer term. Then a lot of things were in confusion there afterwards.

So the local journalists they were called editors?

Yeah.

Did they do any mapping training?

We did, but honestly, I think there was a case of one getting into it, but for the most part they didn't. First of all, they were very busy. They had very, very busy jobs. People always complain about how difficult it is to edit OpenStreetMap. It's gotten a lot better over the years. Back then it was I think more difficult than it is today. We had one editor - one Patch editor, who became an OpenStreetMap editor. The rest of them would send updates to the team and we would do them ourselves or find someone who could make the changes.

And that's a very normal story actually I think, in my experience of getting people involved practically everywhere I have worked. There seems to be an expectation that somehow you are going to get 100% of people to start editing and the reality is you get like 2% to 5%.

In this case, that's probably what the percentage was in terms of how big the editorial staff was.

Why do you think that is?

Well, I think that it has been getting easier to edit. I think it could have been a little bit easier for the types of work people had to do. I think that these things will improve and improve as time goes on.

Things need to be really, really simple. If you have a very busy editorial staff and you also want them to do this, things need to be dead simple. I don't think anything is that simple yet. People would complain about the content management system for example. This is certainly harder than interacting with a normal content management system.

I think there's some kind of conversion rate. It's probably like you said around 2%. I wondered too about FourSquare with Mapbox which has links to edit. I wondered what those percentages are because it would seem like those percentages would be smaller because you are just a casual user at the time. You are not necessarily looking to edit the maps.

What did you do after MapQuest?

After MapQuest, I went to the start up called Brewster, which wasn't in the mapping space, although we did make heavy use of Nominatim. It got me thinking about geocoding, which is a big focus of what we're doing now. It was an address book startup. The idea is you would sync all your social networks. It would take the data and present it back to you through a search driven interface. We were working on that for a while. It's actually still going on. I was the first engineer there and built out the team. Then I left after about two years when I didn't think it was going to succeed as quickly as we would like.

During that time, what was happening with Patch and the MapQuest?

Oh man, that was a real bummer in terms of regrets about leaving. I went to a Ruby meet up in New York, and it was right when Google started charging for their Maps API. I was so bummed that I wasn't still at MapQuest because that would have - when I was seeing this, I think I saw someone whom I didn't know, who wasn't really at the time in the mapping community, but it was some Ruby developer who was saying, "Hey I found this MapQuest API. Google started to charge. We're going to use this instead."

I was like, "Oh man, this would have been amazing." I could have totally done something great with this but I was excited about the new startup that wasn't really mapping related. Yeah, I had always regretted missing out on that opportunity, which I think MapBox has done a really good job capitalizing on. I think that that Google's charging has opened the door for that to become a viable business.

In the cleanest definition of capitalism ever.

Right. The key thing there was having - I was using the Nominatim service. When a user would sign up, they would sync their social networks. They would have - I can't remember what the average was, it's like 2,000 contacts or so. We would need to geocode as fast as possible the locations of your contacts - generally we just did coarse geocoding. Google Maps, I can't remember but they had a very restrictive concurrent request limit. We ended up using MapQuest's Nominatim service which ironically we had set up when I was there. That was very, very flexible. You could throw in any amount of traffic at it. It wasn't quite so limited.

We were able to use that and it worked totally well for our needs which was basically coarse geocoding. Yeah, that was great. We ran into some problems where I would have liked to fix some bugs, but I noticed that it was very difficult for me as mediocre developer to submit a bug fix for Nominatim. Right now, we have a geocoder that we are working on it, taking that as a lesson, to try to make it as easy as possible for any developer, of any skill level hopefully to be able to contribute a fix back.

That's the main relevant lesson that I learned at that startup. That, and search in general. I was working with the ElasticSearch and managing large data assets. It was cool.

What happened afterwards and how did mapzen then come about?

When I left Brewster, it was in April. I really wanted to just take some time off and think about what to do next because it was I had never really thought carefully about what I wanted to do next. I always would have a timeframe and just kind of select the thing that made sense within that timeframe. I took off between April and August, a couple of summers ago, got a puppy which was cool.

Talk about your puppy?

Portuguese water dog, her name is Libby, she's cool. She's like a year and nine months or so now. I always wanted to get a puppy but I wanted to be home for a while, so it worked out.

I just kind of started to get coffee with people professionally for several months. I was taking time just seeing what kind of opportunities there were, what thing can I do for their engineering teams and build products and things like that. I ended up talking to a number of people. David Eun had become an executive at Samsung and he was building this Open Innovation Center including the Samsung Accelerator which in New York City and in San Francisco. The idea was, they were trying to get people involved who had built successful products or companies in the past.

At the time, I was building a fishing app. On the side, I've been always interested in tidal data. I have an app in the Google Play store, in the iTunes store, that's pretty popular, Tides Near Me. I was starting to move this into fishing where you could predict weather events like tides, currents, biometric pressure, et cetera, that would help you figure out if you could catch a fish at a certain place and time. I was just kind of hacking on this.

As I started to talk to the Samsung Accelerator folks I was like, "Yeah, I'm working on fishing app." I'm sure that's not - I just felt that probably wasn't an interesting thing for them, and I wanted to get back into mapping. I started to work out this thesis of what happens when you fully embrace open map data and open software and have it as a rule that you will have no proprietary data whatsoever, which maybe isn't a rational thing to do if you're raising money to build something, but still creates incredible value. If you take a long-term approach and you take the thesis that open data will be proprietary data at some point and you start to put resources around open software and helping communities where they need help and putting resources around that. I started to think about what if we had a data budget like we had at MapQuest. What would it be like to give some of that money to say, Code for America, where the result of their work is sometimes open data.

What if you did things like that for the long-term, could you have a workable consumer mapping application all on open data and software? What we've done is basically just that. We're working on the technical side on a geocoder, a routing engine, and vector rendering. We have a number of SDK's for people to use. We have an app in the Google Play store. It's basically a test app that actually requires OpenStreetMap log in. We're not --

Do you on purpose want a small demographic?

Exactly. I think I told you that a few months ago and you almost spit beer all over the table laughing so hard that we require OSM Oauth. We're just looking for a few hundred engaged users who can give feedback, which we have, which is good. Later on, we're going to be launching other stuff. The key thing is that what we're trying to do is either support current open communities whether it's on software or data or create them where we see that there's a problem.

For example, looking at the geocoder, which is still one of the biggest problems. I think we have talked about this. When I was at State of the Map in Girona, I was standing up on stage, and the back of my mind, it's like, "Will this data ever be navigable? Are you ever going to be able to navigate over open data?" I think the answer basically is yes, at this point, depending on where you are. Now, it's like are you ever going to be able to geocode on this data? I have feeling that the answer will be yes at some point at least in a couple of years, depending on where you are. We're starting to work on the geocoder. We are starting to work on getting more data into the open public domain or OpenStreetMap.

You clearly can navigate over it or at least with what we did with Scout. It just needs a little bit of help to get there.

Yes, I think that the thesis will prove true where this data will be in the public domain, or in the open in some way in the not too distant future. I feel like geocoding is going to follow the same path that navigable data followed.
From Mapzen's perspective what's the major benefit? Is it building devices that help if the software is open and the data is open?

In our case, we get funded. Mapping is clearly a huge strategic area for anyone involved in mobile, whether it's what you saw Amazon doing with Kindle phone or Apple with how they built their own maps or Google is obviously huge. For anyone with any stake in mobile this is a crucial, crucial area. The reason that they would make an investment like this is that yeah, the concept of open mapping if that where to actually work would be beneficial to a lot of people. You could launch an alternative operating system and be able to build a map on it and be able to do it easily and cheaply. That would great for everybody.

For them or for any number of companies, it's a pretty clear reason why you would invest in something like this. It's important to note that with the Samsung Accelerator, this is basically my thesis and the team's thesis that they are backing. Samsung of course works closely working with Google and Nokia on mapping, on the phones that they ship. This is more like a startup the Accelerator is backing.

It's really that large players in the mobile space are erecting these large barriers to entry, like mapping would be one, right? That you can use Android for free is an example but if you want the maps and if you want the Play Store, and its probably other things that are just as important, to make Android functional for the end user. How are you going to negate those barriers to entry long-term? And maps I think is the biggest and hardest area there is. Yeah, so open mapping acceleration is, I think, a good strategic move for any number of companies.

How did you come up with the name Mapzen?

It's funny because Mapzen was a CloudMade product for POI editing I think at some point, but it's long defunct. How did I come up with the name Mapzen? One of the things I was really interested in coming in and working within the Samsung Accelerator was the Tizen operating system which is hosted by the Linux Foundation. Samsung works a lot on it but it's a mobile OS, kind of similar to Firefox OS. The key idea behind it that it's cheap - they're currently selling it as a sub-$100 smartphone in India. I liked that concept. I noticed the Mapzen URL was available for like $500 or some totally reasonable amount, and I just bought it. Then we just had it as a working title for a while, and then it just kind of stuck. It was kind of awkward because, "Wait a minute... People from the OpenStreetMap community might remember this?" My hope was that the people who remember it as something old and defunct will be outnumbered by the people who know it as something new.

The long-term bet is that a lot of proprietary data was going to be opened in some way, shape or form. We might as well accept that in the present and figure out how it's all going to work and build up that expertise and tooling and so on and maps is an obvious big one to go after.

Right, right. That's it. Just expose what the problems are, so if you look at our app, it's a nice app. It works well. It's probably similar to Scout in a lot of ways, a professional, good looking, highly functional app. Except that all the problems of open data are exposed including geocoding, including the fact that you might search for an address and not find it.

For us, given our timeframe that we have to solve these problems, I'd rather expose that and make it really obvious that it needs to be fixed. Let's take the long-term approach and I'll shove this stuff into the open as quickly as possible.

Can you talk about the specifics?

There are services layers, like the geocoder. We have the routing engine. Actually, we've hired the - I think basically, the entire routing team from MapQuest who is working on this, and they're working on a global multimodal routing engine. Meaning, right now, there are number of routing engines that work over open data that are good for walking, biking or driving but not transit. Then there is OpenTripPlanner which works pretty well for all of them except it's only—it's really designed to work in a specific city. The City of Portland was its first use case.

In this case, our team is going to be building something that works globally and works across all of these things, so fully multimodal including transit. That's another service which I think I shows a lot of promise. The thesis behind it is that it's a very difficult thing to do, to be able to offer that as liberally as possible to as many developers as possible, I think brings in a lot of interesting opportunities.

If you look at Waze for example, it's a navigation app but the interesting thing is the social layer that you build on top of it which - it's unfortunate that to build that app you have to have this whole navigation team. Instead we'd like to make it so that app developers can build a lot of interesting stuff if you fully commoditize the navigation layer. That's the thesis there.

Then we have a team working on vector rendering, OpenGL, WebGL across devices and on the web, which is something that I think is pretty interesting. Our team is built of people not really with a traditional cartography background but instead, people from a video game background or animated movie background, or artistic background. They're doing some different things with how you explore maps through 3D space. All in open data which I think is pretty exciting.

Those are the services we're involved in. We'll be wrapping this into SDKs to make them easy to build into apps. They're currently public and being used by some people but this year we'll be promoting that more and getting more people to use this stuff.

Then we'll be building our own apps on top of this. Right now, we have this test app and the Android and the Google Play store but we'll be building other apps that are less restrictive on these tools. We're really focused more on the services than the apps. The apps will be, like almost like demonstrations of what you can do with the services.

Where does that profit motive come in, if it all?

It's down the road. When I was at my last startup, we were backed by venture capital, and what was interesting was the message from investors was, this is an interesting, valuable area, try to get users. This is exactly what you would do with Twitter for example. Get users using the service and then you can figure out how to monetize this. I really like that approach. Don't worry about a business model or monetizing something until you actually have people using it.

Everyone agrees that the mapping space is really valuable. People can agree, especially when you look at the mobile space, that a developer community is really valuable. I think it's more of a longer term thing. I do think that you can create a lot of value in the space by opening things up to people to any number of companies.

I don't think it's a good business to sell map views. I think that's a reality. I'm not telling anyone that that's how we're going to make money here. I think that's a bad business.

Do you think there's any aspect of history repeating itself a little bit?

Yeah, totally. I see it repeating. I think it's repeated at least three or four times in my short career. Yeah, I think the way that the funding works is really important. I think that the way that you try to monetize it or not is really important. I think we're in this place where you see a lot of big players that could benefit from this and it's about providing the right value there for the right cost. Yeah, certainly, history is repeating itself.

When I look at what CloudMade did, that's - as far as I know - the first attempt at a lot of the stuff. There are probably older examples as well. Yeah, at some point, it'll work out. These things are still unsolved problems. There are still people who complain that they can't get a decent geocoder with good terms at a good cost. I haven't really seen - I don't think in the routing/navigation space that it's been made easy enough for developers to actually do innovative stuff there. Part of that is marketing. It's product marketing, really. It's showing what you could possibly do here. It's kind of like the OSM editor. I don't think we've nailed it; it could take decades to get this right.

It could take people from outside of the geo industry to get it right. I don't know. I think it's more of a career-long problem that will manifest itself in a few different companies and a few different places until it really becomes ubiquitous.

It makes me think almost like the great red spot on Jupiter right. That thing has been there for hundreds of years and the individual particles are moving in and out, but the great red spot remains. It's like the problem space remains but it's just different people, different companies coming and going, trying to go attack it.

Totally, totally, and it's a really valuable area now to a number of people, a number of people that are aware of big, big problems with this right now. I'm really interested reading things like the history of the iPhone and how smartphones launched. iPhone comes out in 2007, the first HTC ships, the first Android phone, I think it was in 2008 or so, and then Apple buys its first mapping company and then it works on a map for years. It's like such a key area and it's unsolved. It's not perfect. There's a bunch of people doing duplicate, overlapping efforts. It's not finished yet. It probably won't be finished for most of our career which is good. It's good to have a career.

It's like we're working in typesetting in the 60's or something. I guess 70's, 80s' maybe or some of the industries that are disappearing so fast. I love the phrase, "let me fax my travel agent." It's like, he uses his fax machines. He uses his travel agents. There's so much change that it's lucky to be in an area that's going to—it has its own change, but it's going to be very relevant going forward.

Right, right. Yeah, the change is - yeah, a lot of changes will happen. I think the business model is the thing that will be to go through some change especially.

Over the years, what's the most surprising thing about OpenStreetMap that you've seen?

The community aspect is definitely tricky. I don't know how to tiptoe around the toxicity issue. I feel like it could be a more successful community somehow that could have more data and could have a better community

I definitely appreciate the community attitude on imports. I think Serge actually gave a really good answer about why they're not as valuable - the value in the data is that there's a community behind it and it's not just commoditized bunch of scripted imports which wouldn't be as valuable as actually having an engaged community. I think this makes sense and I'm super sensitive to that. The problem is that we're doing a lot of work to create data now and are not sure what to do with it. Having watched MapBox do a couple of imports of New York City and San Francisco, our take is not to do that at all because I don't - I just don't think it's a good use of company time. It's a stressful proposition, so we're looking at, as we start to create data from different sources, that will be in the public domain somewhere else. That if it trickles into OpenStreetMap, that's great. We're not going to take any lead on importing data into OpenStreetMap. I would encourage the data to be brought in but we wouldn't really do that work ourselves. I don't know if that's a bad thing or not. Maybe not but yeah, it seems like it could be a missed opportunity in a way.

How do you feel about the Netherlands? In the Netherlands, the whole country was effectively deleted and we pulled in AND data and it was actually very successful in the end. We lost people, right, but we also gained an awful lot of people.

I'm not anti-import. I definitely understand both sides. I feel like good data is better. I also understand the community impact and how that's tricky too. Maybe it's just a tricky problem and it makes sense the way the community is right now is not a - it's just a natural progression of how you work through difficult problems like this. It seems like there'll be other projects popping up, like OpenAddresses is a really good example. It has a lot of steam behind it. It's moving along. It's not happening in OpenStreetMap. It's being done by people who traditionally hadn't been involved with OpenStreetMap. I don't know. That seems a little bit like a missed opportunity there.

I feel like you might have already answered this but what's the one thing you would improve in OpenStreetMap?

I think that it's some way to maybe to put data layers on top of each other so it's still accessible. For example, we're pulling all these address and POI data out of the Common Crawl that we want to publish. I'm not going to consider OpenStreetMap, but it would be great if I could do that and do it in a way that it didn't harm anything or it didn't harm any community. Maybe it's different layers or projects somehow.

Randy Meech

I think some way to accommodate both aspects like imports and the community would be really good. You're going to see these datasets accelerate elsewhere and like it would be nice if there were mechanism for all to be in the same place.

Ed Parsons

Bio

Ed Parsons is the Geospatial Technologist of Google, with responsibility for evangelising Google's mission to organise the world's information using geography. In this role he maintains links with Universities, Research and Standards Organisations which are involved in the development of Geospatial Technology. He is currently co-chair of the W3C/OGC Spatial Data on the Web Working Group.

Ed is based in Google's London office, and anywhere else he can plug in his laptop.

Ed was the first Chief Technology Officer in the 200-year-old history of Ordnance Survey, and was instrumental in moving the focus of the organisation from mapping to Geographical Information. He came to the Ordnance Survey from Autodesk, where he was EMEA Applications Manager for the Geographical Information Systems (GIS) Division.

He earned a Masters degree in Applied Remote Sensing from Cranfield Institute of Technology and holds a Honorary Doctorate in Science from Kingston University, London and is a fellow of the Royal Geographical Society.

Ed is married with two children and lives in South West London.

Favorite Map?

"As for my favourite OSM map, it is actually a poster that Steve produced in 2005 of tracks of London Taxi's which clearly showed not only the road network of the city, but also which roads were most popular.. It's was for me a real insight into what OSM could become, and it was talking point as I had it on my office wall at the OS!"

Interview

Who is Ed Parsons and what were you doing before you found Open Street Map?

What was I doing before? Well, OpenStreetMap I think happened bang in the middle of my time at Ordnance Survey. I was then the Chief Technology Officer at Ordnance Survey—the first and only Chief Technology Officer at Ordnance Survey.

They broke the mold.

I broke the mold and they'll never employ someone like me again. I think, it came at a really interesting time. We'd been working hard at OS trying to move things away from digitizing paper sheets that were so big into digital files that were so big and then trying to have a more holistic view saying, "Well, this is spatial information. It's going to be used by people on their mobile phones. It's going to be on refrigerators. It's going to be everywhere." The kind of the stuff that we talk about now, we were imagining that might happen and we were trying to move the organization along. And then up came these upstarts who go running around making maps of the world. There was a real fear actually at the Ordnance Survey that "we don't know what these guys are doing. We're losing control." And there was an element that saw the OpenStreetMap Community as kind of rebellious.

This was the right timing. There was the mapping party at the Isle of Wight. It so happened that Vanessa, the CEO at the time, was also going to be on the Isle of Wight at the same time you were doing the mapping party. She was petrified that you guys were going to turn up and kind of demonstrate by surveying or something. But it was at an odd time. There was this kind of real fear of "What is OpenStreetMap? What are they doing?" And I think I met you just before that, at a conference in Cambridge, I think. Do you remember that?

I got joking to you over a beer, "There's really something to this. Why not, you know?" People are carrying around GPS now, recording tracks, Mechanical Turk was popping up and people were looking at crowdsourcing information so I said "Why not? Why not be able to produce a map in these ways?" In Ordnance Survey's terms, we were looking at making kind of small incremental changes to make the surveying process more efficient. And then along comes the OpenStreetMap community with a completely different approach in many ways. That's what I was doing before. I was seriously into GIS, Oracle databases and all of that stuff and then along come you hippies and saying, "No, no, we're going to do it differently."

Can you talk about how the process was different?

The Ordnance Survey had been capturing largely the same specification of data for all 50 or 60 years. Despite the fluffy reputation and view of the Ordnance Survey as ramblers and so on and so forth, much of what drives Ordnance Survey still is security concerns. The specification is still largely driven by Defence requirements. When you go out and you survey a wall as an Ordnance Survey surveyor, you survey it as an obstructing feature—something you can't shoot through. The guys were going out doing surveying in their own local neighborhoods based on a big thick book saying that "These are the classes of content that you go and collect information from." We were trying to make that process more automated.

It's a complex process. You need the full transaction history. You need to be able to do long transactions, check data of your database, go out in the field with a tablet, update the data, then re-submit the data and make sure that all the changes are tracked and are relevant, and you maintain topology and all of that difficult

GIS stuff. So, in some ways, it's not that different in that you're capturing coordinates but there's much more process involved and there's much more of a complex and very strict data model but, fundamentally, it's the same thing.

What were you hired to do at OS? Was there a difference between that and what you actually did?

Yeah, when I started, there was a very distinct feeling that very, very soon the OS was going to be privatized. It's amazing how these things are cyclical. And so, there was a lot of focus on meeting the needs of a broad community of users, both in the private and the public sector. More focus actually on the private sector – identifying partners, potentially looking at producing products that the customer would use directly either on their mobile phone or—

There was a lot of focus on customized mapping, just print out an A4 sheet of the walk that you want to take this weekend and we'd charge you 50p for it. Can we build infrastructure around you doing that? Could we build, for example, kiosks in visitor centers in national parks that just print out maps on demand for people to use? So there was quite a broad scope in terms of the things that the Ordnance Survey was going to do when I was hired.

At some point, I guess a year or 18 months later, the focus changed and it was much more about the public sector and serving the needs of other government customers. MasterMap became the major project – the thing that everyone was working which was the kind of database version of the old Land-Line cartographic product. And so a lot of that broader focus on getting geospatial data out into the hands of the consumer faded and it became much more of, I suppose, a traditional customer base of the public sector, utilities companies and professional users. It wasn't, from my point of view, the best decision but went along with it because it was still the big task—a big job to do.

It sounds like you were hired to check the box of being a CTO but then you made the mistake of actually doing things?

Yeah. I think, actually, what they wanted was an IT director but they thought they'd hire a CTO because that sounded like the more hip version of what an IT director was, but not actually understanding what CTO job description was.

Let's talk about that conference that was at OS with what's his name, he used to be head of BT Research. Cochrane.

Cochrane, yeah. That was interesting. There was always at Ordnance Survey and still is a very strong group that are focused on innovation, on doing things differently and not necessarily being on message or being completely in alignment with the business model of the organization at any particular point in time. So what you want from a group that does innovation is to ask awkward, difficult questions. That group organized this kind of conference that we should have brought in people from a broader community, people who were thinking about the future. And I think that was—for many people at the Ordnance Survey, the first time that they got a view that actually there was a whole new community developing many of the same things that the Ordnance Survey were doing but from a different technology platform.

Remember the *Mapping Hacks* book that was out at the time. That was quite influential. I gave copies of that to almost everybody who was working for me, saying, "Look at this. This is how our world is changing. Go and play with some of the recipes in here." It was a really fascinating time. And I think, we forget today just how much change and how dynamic things were over that period of a year or two years where this kind of whole mapping hacks, the Google Maps API, and all these things suddenly appeared and made much of what was traditionally a very difficult thing to get into, suddenly very open.

So you almost had like an education mission internally?

Yeah, I mean, not in a sort of idealized point of view. It was more to say, "Well, the market has shifted and these are things that we need to be aware of." Not in any point ever suggesting that "Hey, we need to stop what we're doing and just go and do things differently," but to be aware that actually there were other approaches that we could take. There was a community of potentially very interested, enthusiastic guys outside the Ordnance Survey who would quite like to be part of the process. Like, "Could we bring elements of what the OpenStreetMap community was doing inside Ordnance Survey to help in what was and still is Ordnance Survey's greatest problem?" which is about intelligence, finding out where something has changed so you can go and survey it. And that still is what OS spends most of its time doing.

I was just wondering if you remember what Cochrane said to the audience and the shouting match afterwards.

No, remind me.

After I did the talk on Open Street Map, he said something like, "This is the future. We just have to wait for you guys to die." And people were very upset. He's like, "Look, we don't want your opinions. We just have to wait for you to die and leave." And people were like, "You're saying I'm going to die." And just very odd audience reaction.

Well, I mean, I guess, that was him doing his job. You pay for this great technology guy to come in. He's got to try and ruffle some feathers but there was an element of truth in what he was saying—the OS isn't dead yet. But clearly, they and like all mapping agencies around the world are kind of having to re-assess what they do, not so much from a producer point of view but I think from the fact that a lot of the users, who saw mapping agencies as the only source of mapping data, are now all saying, "Well, we can go to OpenStreetMap. We can go to Google Maps. We can just be happy with the little dot on our phone. And 78% of our needs are delivered by these other sources of information. When we have very particular needs for authoritative, very detailed information, then we'll come to you mapping agency. But actually, for most of our needs, we don't need you." And that's still an issue, so a lot of mapping agencies do get over.

Were there other things, while you were there, that the OS was concerned about? I mean, did the iPhone come out at the same time or anything like that?

No. In fact, the iPhone came out just as I had left actually. But, again, there was mapping on phones before the iPhone. It's funny quickly we forget. There were the Nokia phones. The ones that were running the Symbian Operating System. They had maps on them. The Palm phones had maps on them. We could see it coming. Multimap was big in the U.K. at the time and Google Maps had started. There was certainly the awareness that there was this shift of gravity more towards consumer applications but it was nowhere near as big as the shift is now. It was starting. You could see it happening but we were still—the OS was still largely meeting the needs of the professional market.

I've forgotten all about Multimap. I'm going to add that.

Yeah, Sean Phelan will be sitting on a boat somewhere in the Caribbean saying, "You should remember what I did."

Yeah, he made all his money off Microsoft and then started to do eco-investing, I think.

Yeah. Oh, he's still doing quite well. I mean he did a great job. He grew it from— he was one of those great British startups that people forget. He and Audrey had the server running in their back bedroom for the first six months of the year over the history of Multimap.

Okay, so you left OS and then what happened?

I left OS. It was picked up in The Guardian. The very next day Google, in London, phoned me up and Michael Jones, from Google in California said, "You need to come and work for us." And when Google phones, you don't turn it down. I was very excited that Google, at the time—I hoped, and it has been the case, has done a huge amount to get mapping and geography in the hands of the world. That's a fantastic opportunity so, of course. I still had to go through the arcane hiring process of being interviewed by a dozen people and you never quite know what's going on and it takes a very long time but it was great to be head hunted, for sure.

Did they ask you to solve math puzzles at a white board?

They did, yeah. They were quite specific map problems. They were saying, "Okay, given I want to represent the world with 30-metre pixels, how much data storage do I need? And how can I reduce the amount of storage?" It was quite practical, not particularly abstract. But yes, you do have those questions asked.

Okay and what was happening at OpenStreetMap at the time?

I think, at the time, OpenStreetMap was kind of expanding from the U.K. which had probably just been completed, or was in the process of almost being complete then. I think there were many of the debates starting to knock around about "Should you import data? And what's the best methodology of really scaling this up?" It was still, I guess, kind of relatively early days.

Do you have any memories of different conferences you went to?

The conferences I had the best memories for was the conference that was in—was it Manchester, when I was still at the OS. It always seems to be that when I go to an OpenStreetMap Conference, it's like, "Oh, here comes the enemy." Regardless of who the enemy is at that particular point in time, I represent them. Once I was there representing Ordnance Survey as the enemy.

And then I remember the conference held in Ireland and Google had just announced Map Maker. It was like Map Maker was announced on the Wednesday and the Open Street Map Conference was that weekend. And I had planned, being the great aviation enthusiast that I am, to go to this big air show that's held at Fairford every year. I was all set to go to that and thought, "No, I really do need to keep driving west and get on the ferry and go and talk to these guys because you all get the wrong end of the stick." Because, at the time and ever since, I've always maintained the fact that Map Maker and OpenStreetMap were different projects. They were trying to do very different things. I didn't want people to perceive that Google was kind of trying to jump onto the OpenStreetMap project and destroy it by branding, marketing power and so on.

For someone reading this book, why do you think people perceived you as the enemy?

Well, I think, it's just a scale thing. I think that OpenStreetMap had been very successful at that point and it was somewhat growing in strength day by day and was beginning to be seen. I suppose, yeah. I guess, it's always been seen as the kind of response to the man. The man being whoever is industry leader—big, in control.

Clearly, I think, it could be the perception that Google appears and will want to try and dominate this process and say, "Okay. Yes, you guys, that's a great idea but we're here now to take over." That was never the rationale behind Map Maker but you could easily perceive it that way. And if you wanted to portray it as "Here comes big nasty Google," it would be quite easy to do that.

And so, I tried in my presentation that Open Street Map—and States of the Map and the other occasions afterwards to say, "Look, Map Maker is all about getting Google Maps up to date." It's not about exporting data to whatever other application. We have quite restrictive terms of service around how the data is going to be used. It meets the needs of our user community as opposed to enthusiasts – people wanting to develop new applications, people wanting to innovate because that's a different community that is served very well by OpenStreetMap. The reason why Google have always supported OpenStreetMap is because of that. It's because when you have that open access to data, you can do really interesting stuff with it. It's hard to do the really interesting stuff when you're always looking over your shoulder to try and work out, "Can I do this, under these circumstances?"

How was that received?

Not greatly. I don't know. You were there. You were in the audience, what do you think? Did anyone believe me?

I honestly don't remember.

Subsequent to that, I think, it's always been—it's been convenient to see Google as the big guy and it's nice to be the rebels you know, in OpenStreetMap. Fine, if that's the perception you want to continue with, but that's never been Google's view.

Do you have any memories of any other conferences like Manchester?

Well, these conferences, they all kind of fade into each other, don't they?

Yeah.

It's likely because you're having such a great time. The thing that I always love about the State of the Map Conference is the fact that you've got such a diverse bunch of people. They're all driven by wanting to create something, wanting to add to what the community does, add to the information and knowledge about their own particular neighborhood which is interesting—that automatically kind of self-selects a group of people. But then, once you've kind of got over that, then it's a very diverse group. You've got the real kind of geeks who are out there and talking about XML tagging schemas. And then you've got the guys who just quite like walking and carrying a GPS in their backpack and exploring the world.

I think what's perhaps changed over time is that you've got more of the IT developer—perhaps even more of the entrepreneurship has come into it where you've got groups now—companies that are making a living building solutions on top of OpenStreetMap. And they're kind of a different community that gets involved in the conferences. That's progress. That's how things develop. But I think sometimes as a result of that, there might be a bit of a tension between the guys that just enjoy wandering around making maps and the guys that say, "Well, no, if we're really going to be successful, we need to have proper network topology or we need complete address ranges. If we're going to make a product from this, these are the things that we need to do."

I remember, we had early discussions of—in the early days of Open Street Map, I remember saying to you that at some point, there's going to be a need to be a bit more direction to say, "Okay, yes. We've covered most of the landscape but we now need to focus on these things." It's because you can't create a project as complex and as sophisticated as a digital map of the world and assume that you can build it completely from the ground up with no design. Sometimes, you do need to take a bit more control and say, "Well, oh that's great but now we need to concentrate on these elements that are important but they're missing because perhaps they're boring or they're a bit dull but nevertheless we need to do the work."

I'm aware, yup.

And it was always going to be a struggle. It was always going to be hard motivating a community to perhaps do things that aren't as much fun. Everyone has to write the printer driver to the operating system but nobody really wants to do it.

Yup, that's what I used to tell people I did at Microsoft. I worked on the printer drivers. Okay, so then, somewhere in here you're being recruited by Google. What were you recruited by Google to do?

Largely, to be outside of Google. Google is an organization. It's still very much engineering focused, lots of engineers going—working very studiously, doing clever and neat stuff and seldom having time to stick their head above the parapet and find out what's going on in the rest of the world or to communicate clearly, succinctly what they're doing to the rest of the world. My job was to be more of a public face of what Google does.

Initially that focus was around Geo. It's gotten broader over time just as, I guess, the scope of Geo has gotten broader. But it was to be that kind of the guy sticking his head out saying, "Okay, this is what we're doing" and also to try and work out what was going on, on the outside and feed that information back. And I think, even today, we probably still struggle at communicating what we're trying to do. When we really mess up and be criticized is often because we haven't clearly communicated what we were trying to do and people have misunderstood or misrepresented what we were doing.

Do you want an opportunity here to clean up anything?

No. I think that when we make mistakes, they've been real mistakes. There isn't an evil plan. Sometimes shit happens and you just have to deal with that.

Forgive me for jumping back but what was your reaction when you very first saw OpenStreetMap?

I guess, the first time I saw it was that early trace, I think, of the taxis in London. My first view of Open Street Map was just London. It was "here we are building this map of London." And I thought, "Well, that's cool because we can see the street network."

The thing that struck me most though was that immediate insight that you can actually see what were the most popular roads because the GPS wasn't that accurate. You had thicker lines where people were driving their taxis up and down Piccadilly. These roads were much thicker than the other roads and immediately you can say, "Well, you're not just capturing the geometry here. You're capturing the flow of information. You're capturing a view of the city as a sort of living entity and that kind of registered in the back of my mind saying, "Okay, there's more to it than just creating a map. There's more to crowdsourcing content that is coming from people carrying out their day-to-day lives but with a device that's portable. There's much more value in this than that kind of very first view of that map. It was like, "Ooh, yeah, there's something interesting here." To be honest, I thought at the time, "Well, that's great but it's going to be hard work turning that into a proper street map with names and all the rest of it" because—

I think in the very early days a lot of the focus was on doing those GPS traces and potentially kind of tagging your content as you went along." I remember you turned up with your backpack which had a big beefy laptop in it and a GPS connected to it. It was kind of hard technology. I guess, we didn't appreciate at the time how important the community, just sitting in that room, tracing imagery and editing content based on their own kind of neighborhood experience would be, that's where the real growth-the real value of has come from. Not with people running around with GPS in their backpacks but people sitting in front of their computers. So when people say, "What was the technology that allowed OpenStreetMap to be successful and to grow?" It wasn't GPS. It was the internet. It was a browser. It was allowing people to edit a common database. That's what made the difference.

Good because you're focusing on the piece that I built so I want to talk about that although I was expecting you'd say aerial imagery actually.

Well, I mean, obviously but I think that's part of the solution but it's still that local knowledge. The aerial imagery allows you to trace geometry that's not there. But the stuff that's really difficult to obtain is the fact that this street is this street but only over this length of this particular segment of road and then they change it to something else. It's those kinds of nuances of the local geography and the kind of landmark features that are important that only local expertise brings. The fact that this is then exposed by OpenStreetMap, that is so powerful. Now you can automate the process of creating a database of street center lines. There's value in that but not huge value. The value comes from the attributes that you attach to that street network.

Can you talk about the difference between how OSM was seen at OS and how OSM was seen at Google?

There were some similarities. I think, at OS it was always seen as "Oh well, it's kind of Mickey Mousey. It's not going to threaten ultimately whether the main business of OS was"—which is large scale topographic data, once Land-Line - now MasterMap. There wasn't a perception of not being threatened. Potentially, the smaller scale business geographics data as it was called, was threatened.

Likewise, at Google, I don't think OpenStreetMap was ever seen as a great threat. It was seen as an interesting technological innovation. It was seen as an interesting experiment in community mapping and in community process. It's hard, there is no one Google view so this is kind of my perception. Google has as many views as there are people working in Google. I think it was seen generally much more as "That's really interesting. We need to watch that. We need to understand how things are happening." It's a great, almost laboratory experiment to see how a community can create a map of the world. We can recognize its strengths but we can also see its weaknesses. So, it was something that we watched and we continue to watch.

Do you want to talk about how Google has supported OpenStreetMap over time?

Well, I mean, it's largely being financial. That's the easiest way to support you guys. We've done supporting conferences. We helped financially with some of the servers early on. We've always supported through the Summer of Code Program people wanting to build new tools for various editing platforms over time. It's been, I suppose, a little bit behind the scenes but it always has been there.

I guess, we've probably fallen off a little bit in the last few years. But regularly, I think, we support dozens of Summer of Code Projects that are around OpenStreetMap. And that's, as I said where our interest are. It's kind of building this ecosystem around the use of geospatial content whether it's geospatial content that is OpenStreetMap or another source. It's interesting for us because Google benefits when the internet gets bigger. And if that's us growing the internet or other projects growing the internet, great. The bigger it gets, the better it is.

And so over time how was Map Maker? First of all, what was the purpose of Map Maker which you talked a little bit about? And then, how was it implemented? What are the differences between it and Open Street Map? How has it evolved over time?

Map Maker was always designed to keep Google Maps current—up-to-date—and in the early days it was created to allow us to fill in content for those parts of the world where there was no other mapping data because remember in the early days of Google Maps we were licensing data from TeleAtlas and wherever or what other sources of information that we could get. It just so happens that TeleAtlas didn't have data for certain areas because there wasn't a market for it in navigation systems, so the content just didn't exist—so India—there was no mapping available. Map Maker, as a tool, was developed to allow a community of users in those countries – India, Africa, lots of the Far East, to start to contribute maps of their own local areas by tracing on top of Google Earth satellite imagery made available in a browser.

It was, I suppose, most different to OpenStreetMap in that we had one single editing interface onto a very fixed data model, that data model representing the content that we had in Google Map. So we had a particular set of classifications for roads and highways. We had particular types of points of interests, particular ways of structuring the content. Now, users were filling in that information as they found it.

It allowed us very, very quickly to produce a global map of the world. And then, ultimately, above and beyond that, to then keep and maintain a map of even the U.K., the United States, and Canada. Everyone's was soon building on Google content—on a Google map of the world. It was quite fundamental to Google being able to really deliver on this idea of having universal access to mapping. Universal access to mapping means wherever you live, there should be mapping content for your area available to you on your mobile device, onto your laptop or whatever so Map Maker was pretty crucial in delivering on that.

So why not just use OSM?

Because of the license, Steve.

What were the issues with the license?

Well, the issues with the license were issues of clarity at one level and the viral nature of the license being the old CC By-SA license or even the up-to-date license. There is still that viral element that "Yes, it's open to interpretation exactly how viral that is." But if it's open to interpretation, it means our lawyers will interpret it and that causes problems. And so, it was that viral issue that's always been the problem. If we make an edit, does that edit have to find its way into our competitor's products? Maybe not, but do you find the point in it?

Right. Where do I go from here?

This must be something you're well used to discussing, Steve.

It is but I'm usually on the other side of the table. What do you think about the idea that as OpenStreetMap grows, it sort of floats all the boats and it's not necessarily a competitive issue, that we're not really

competing over the base map data. Of course Google is, right? But the rest of the universe that's just using NAVTEQ or TeleAtlas isn't competing over the base map, they're competing over other stuff on top or at least attempting.

It's hard though, isn't it? It's hard to separate the base map from the things that you might add value to. If the base map was just road center lines and military charts of world, digital charts of the world, 1:1,000,000 scale. You have to get most of the roads, that's what we need. That will be fine. But everyone knows that's not the case.

And there is no such thing as a simple base map of which you can then pick and choose the attributes that you add. The reality is, to deliver most of the product and services that we expect, you need a network, topologically structured database with turn restrictions. You need address ranges. You need points of interest attached to that road network. You need intelligence.

And that intelligence is already in OpenStreetMap. It's being built into Open Street Map. It's built into most of the databases that sit behind most of the services that you are using on your mobile phones. And all of the commercial companies involved have invested huge sums of money in building that intelligent map so it's highly competitive.

Do you want to talk about the things that Google has built on top and I'm thinking street view amongst other things?

Well, Street View started off as a sort of an experiment. It was Larry and Sergey driving around with a digital SLR sticking out their car. But from that recognition, they weren't cartographers. They weren't used to using maps any more than any other normal person. And they perceived that weakness that we have to recognize that when you zoom all the way in and you're looking at a section of the street, all you're seeing is a line or potentially you're seeing a line with a few building outlines and you don't get that sense of place in the same way that you do if you've got an image.

And Street View, for us, was always that recognition that this is the ultimate level of zoom. You can zoom all the way in from space and you can then be standing on the street corner and there's huge value in that. It's really—despite some of the privacy issues that we've had, it's by far one of the most popular features that we have in Google Maps. It's one of the reasons that kind of Google Maps is so popular is because you can zoom all the way in and then you're standing there looking at the restaurant you want to visit or the office that you need to go and/or you're looking at your house.

Do you want to talk about the privacy concerns and how they're fixed because not everyone is going to understand that?

Yes. When we started off in the States, the resolution of imagery in Street View was less than it is today but we didn't perceive it to be a requirement to manipulate that imagery so there was no blurring of people's faces or car registration plates. And indeed, Street View wasn't the first time that people had tried to get street-level photography. There was the Amazon project that had been going on a few years beforehand. There was the Map of Boulder that had been created in the '70s. All the way up until that point, there'd never really been any perceived issue because we were capturing imagery as you would see if you were standing on the streets and you don't need to blur photographs.

It was only when we started Street View in Europe that people began to be concerned. And that's because, I think, there is just the different view of privacy in Europe. The definition of what's personal data in Europe is much broader than it is in the U.S. Someone who is absolutely recognizable is personal data, so we had to quite rapidly implement a process that will automatically blur anybody's faces, automatically blur any registration plates. Then above and beyond that, to provide a mechanism for people to say, "Okay, even though you've blurred my face, you've blurred my car's registration plate, I still want you to remove this section of Street View from the database." You don't have to give a reason, we'll just remove it. To be honest, it's kind of much less of an issue now than it was a few years ago but those were some of the issues that we had to go through. And then of course, the biggest SNAFU was when we accidentally picked up loads of people's Wi-Fi traffic—not a good move.

When I used to live in Reigate, I took a photo and it used to be the primary photo for Reigate Train Station on Wikipedia. Can you spot the error? So the person who owned the car in the photo started contacting me over Wikipedia, like messaging me through the messaging system in MediaWiki because their friends had alerted them that a picture of them illegally parked in front of the station is now on Wikipedia. They sent me all kinds of messages trying to get me to take down the image. I think my response at the time was, why don't you go and take another photo and replace it, right?

Yeah, fair enough. But it's fascinating, you suddenly discover that the world is this colorful tapestry of strange things happening. And when you get a picture of the world and if you spend time looking at it you'll find all sorts of strange things. That's reality. Street View wasn't the first time that this was done but it was the first time it was done globally or at that scale. And that's why it became such a, I guess, an issue but much less though now.

Do you want to talk a little bit about society or governmental response in general? I'm thinking this bananas law that may be forgotten and all these websites now with the stupid f*ing banners that say "We use cookies," right?**

We use cookies. Some people still don't know what cookies are but they know that they're in use on websites. I think, it's just a stage of going through. Technology changes just so rapid and societal change and in particular political change just can't keep up. You then get those disconnects where quite valid concerns around privacy and around how data is used and shared are absolutely valid concerns for politicians and for people to have. It's just that sometimes perhaps kind of a knee-jerk reaction to do something isn't the most appropriate long-term solution. But that's kind of okay. We go through it. We get it. We get the fact that we have to co-respond to legal frameworks in different countries and do things differently.

I think, in the long term, I hope that a balance is reached where peoples' privacy, peoples' data is protected while at the same time we don't have a man with a red flag walking in front of our car because there's a perception that it's dangerous. We're kind of at the man with the red flag stage of the internet. We have now reached the point that this is important and valid and of enough impact to society that government thinks that society needs to be protected in some way. And indeed, society does need to be protected. We just haven't quite worked out what the best mechanism is.

I'm a bit more cynical, I think. In the, U.S. it's illegal to make a toothbrush without FDA approval. In the State of Florida or if you're an interior decorator you need a license. There's lot of Federal talk about regulating the internet to preserve net neutrality – whatever that is. I'm trying to make this read okay in the transcript so I'll just be frank. I think the government's figuring out how to get its claws in on how to extract money just like they do with all the other industries, right? It's just in the beginning, yes, anyone can build a car. Then you did stupid things like guys had to wave a red flag in front of it. But in the end, I mean, it's virtually impossible to make a car because there's so many rules and regulations.

But we don't have the man with the red flag anymore. I think you're right. That's what part of what government does. It's that it wakes up and says, "Well, society expects us to manage." I'm probably as much a libertarian as you are in these things but government still has this role to manage and protect and it does. It takes a while to really understand a technology and to understand what the impact is going to be. But ultimately, these technologies do grow. If you're flying around in your Tesla, the government hasn't had a huge impact in stopping you from driving around in a huge piece of technology that's amazing but you could still—

They bailed out GM, Ed. I mean, they could've taken that money and done something else, right?

Yeah but, I guess, that's when governments do what they say they're not going to do and then interact with the market but, ultimately, you can still trace from your Tesla all the way back to Mr. Benz and his horseless carriage. It's just that in the early days, the constraints were perhaps more constraining than they needed to be. In the early days, airlines were government agencies, you know? The early days of Lufthansa, Imperial Airways, KLM – they were all government organizations. Private enterprise took over the airline industry.

If you look at the history of all those things, it's because for example the railways are a much better example because they were so much bigger at the time. The railways were private enterprises and they got regulated to death – to the point where they weren't allowed to raise ticket prices. With less money, they had to skimp on maintenance, so they became really bad at running trains to the point where the government nationalized them as a result of their own policies restricting their income in the first place. And now they're being put back in the private sector slowly but surely across the world. I'm trying to say they weren't national entities because that was the best thing to do. It was because they were screwing everything up, right?

Yeah. These things are often cyclical. What starts off as private, becomes government, becomes private again. Your government will always get involved. I guess we'll always try to do so largely from the positive. Largely, to try and make things better but often don't make such a great job of it.

Last question along this line. I'll try and make sure I translate this the right way, I think one of Peter Thiel's points in Zero to One would be that the reason that technology is doing so well, and is so profitable, and is producing so many interesting things in the world is precisely because it's not strangled by regulation. Because if I wanted to build a skyscraper, it's basically illegal. If I wanted to build a car, it's basically illegal. But if I want to go build a new website, I have zero interference, right? Would you agree with that?

I would. I think that the great growth, the flowering of economic and cultural value that has come from the internet has largely been because it's been unregulated and you don't have to ask anyone to build a website I think that's great. But at the same time, you have to recognize the fact that for every good use of the internet, there are bad ones and the nasty stuff that's going on is unfortunately is what gets reported in the newspapers. You don't see it reported in the Daily Mail, "Look at this great company that's now, exporting hundreds of millions of pounds worth of their product or they're designing these great clothes and selling them around the world because of the internet". No, you hear about the predators and you hear about people slagging each other off on Twitter. That's just the way the press works and then the politicians, as always, will respond to what the press is talking about.

So what has been the most surprising thing you've seen from OpenStreetMap over the whole ten years?

I think the thing that's been most surprising is the scale of the endeavor and recognizing just how many people are interested in what is quite an arcane and difficult to explain process. OpenStreetMap is still quite difficult to get into and get your head around from the technology point of view. And also, to kind of connect into the community in the way the community works.

Given that the number of people that are involved – Okay, not everyone's active but all of those people are registered. All of those people take part. All those people are aware and knowledgeable of it. That's been amazing. That scale, that number of people that are interested in the world around them and putting that world into a database. I don't think anyone would have ever expected that.

I remember, in our early discussions, our scope of this was limited. Do you think we'll ever be able to get a map of U.K.? Yeah, probably. Could we ever imagine that you'd have the most detailed map of the planet? No.

Funny how it turns out.

Yeah but it's that scale thing that's amazing. Absolutely incredible. I guess that's the network at work. That's what the internet is all about. It's the biggest lever on the planet. You give me a lever and I can move the world. Well, give me a network of people that are enthusiastic and I can really do incredible things.

How do you think OSM has influenced? And take your pick whatever it's influenced, Google, OS, society, mapmaking, GIS or whatever you want.

I think it's worked at two levels. I think it's worked at kind of a technology level saying, "Yeah, there is something to building platforms that allow users to contribute content." And not just contribute content but to do that in a managed sense because it's more complicated than just uploading a YouTube video of a kitten falling off the sofa. You've got metadata. You've got to understand data and how your piece fits in with everybody else's piece. But nevertheless, you've this big community doing that. So from a sort of a technology platform point of view I think more than maybe even Wikipedia. It kind of demonstrated that that's possible. That something is doable and has impacts in all sorts of different markets.

The other thing, I think, is it's had a huge impact in making mapping and geography relevant to the man on the street, not in any way as much as Google has, I'd like to say, but that it's still making it into something that's accessible.

I hate the democratized term but it has. It means that people feel that maps aren't something that is designed by somebody else. It's something that they have a role in contributing to. The maps were always something that you were given. As a citizen, you were given a map that a commercial company or your government gave you.

Now, you can create your own. In the same way that we can't all be great successful novelists just because we've got a word processor, we can't all create detailed utility maps. But for representing the world around us in a way that is closer to our world view, that's a very powerful tool. It means that maps are owned by the people that create them. That's very different.

What is the issue with the term *democratized*?

I think it's just over used. I think it's lost its meaning because of its overuse. Actually, I think, in terms of what OpenStreetMap actually is about is probably one of the purest meanings of being democratized because it truly is now a democratic process. I, as a citizen, can help contribute to the map of country X, Y and Z where in the past that wasn't something I was involved in. I'd argue that's quite a pure and valid use of the term. It's used in all sorts of other places to for example democratize video production. Well, who cares? Since when was making a video part of the democratic process? Making a map of where you live is absolutely part of the democratic process.

That makes sense. Where do you see OSM and anything else e.g. Google Maps going in the future?

I think OSM is kind of going through those awkward teenage years. I think, there is that tension. You know this more than anyone. I think there is that natural conversation now between people wanting to professionalize, to better structure OpenStreetMap with a view as to how it's used. And the community, they just quite like making the map and are not that bothered about how it might be used or who might use it.

I don't know how it's going to work out. OSM ultimately had a big fork. Does it have little forks? Does it carry on as it's carrying on now? I don't know but I think this kind of fundamental conflict of being user-focused or producer-focused needs to get solved because it's hard to keep the two together. From a Google point of view, Google Maps is all about adding a geographical context to information. Now, most people interact through Google Maps via their phone—not necessarily directly with the maps app. People are interacting with other technology that we've now built through things like Google Now. Or if they're using Inbox to say, "Well, I want my e-mail to pop up when I get home." You know, it's become so much part of the underlying infrastructure now that you need to speak to a digital database of mapping data to even make an e-mail application work.

That's amazing. I love that, that it's become so embedded. Now, I think, it's as important now as your laptop having a clock. You don't think about all your computers having a clock but the clock's used in almost every function that you have in your computer - listing your files, looking through your e-mail in the same way this geospatial technology is going to be so embedded in all the functions that we carry out in our day-to-day lives with interacting with the internet or our own data on our devices that we almost forget that it's there.

Yeah, my favorite example is spell check. That used to be like a paid feature and now it's just in every text box.

Yeah and I think that's great. It might mean that, as professionals and as the technology—we become a little bit less visible. But that's okay, we've still got a lot of hard work to do behind the scenes to make all of this stuff work. I don't mind the fact that there might not be GIS conferences any more. Well, probably there aren't really that many GIS conferences any more. We've gone through that stage.

What's the topic that we haven't talked about?

You, Steve.

Let's talk about Steve. Go. Lead on, Ed.

No, I just kind of want to make the point that I think it's impossible not to recognize the role that you've had in making all of this happen. Yes, it's been a huge community and lots and lots of people have been involved in the process up until this point. But I think it needs to be put on the record that none of this would've happened without you being an awkward bugger and saying, "No. Buggery, I want a map of London to do whatever I want to do and I'm not going to put up with the fact that OS won't give it to me. Lots of people would have said, "No. Well, never mind." And turn around and try something else. But you, being the awkward get to—"No, I'm going to try something." It's been hard work all the way through that a lot of it has been on your shoulders. So kudos to you. Largely, it's still your project.

Well, thank you. It's funny because most of the people I talk to in OpenStreetMap have the opposite point of view.

Yeah, but they don't have the perspective. That's the thing. But you can't deny the fact that you created this. You were there at the beginning. Somebody else could've done. And probably someone else would have done but they didn't. You did.

It's very entertaining. Some of the people I talked to about this, they literally, their job title is OpenStreetMap Engineer or something and they send me e-mail about how I'm irrelevant to the project and didn't really do much in the beginning. I'm just like, "It's okay. Thank you."

Yeah. Well, you can't airbrush history to that extent. You can't deny the role that you've had. Maybe five years, you're off running whatever business you're running. You might not have any involvement. That's okay but you can't deny that at the beginning, it was your project. It was your baby.

Andy Robinson

Bio

Andy Robinson joined OpenStreetMap as a volunteer map data contributor in September 2005 going on to develop the initial ìMap Featuresî data tagging schema and serving three years on the OpenStreetMap Foundation board. Andy lives in the UK and works is a civil engineer on underground projects around the globe, which has allowed him to map many specific locations as well as extensively in and around his base in Birmingham, UK.

Andy was instrumental in starting Mappa-Mercia.org, a local OpenStreetMap focus group for the UK West Midlands, which promotes the project and organises regular social and mapping events.

Favorite Map?

"My favourite part of OpenStreetMap is the National Memorial Arboretum in Alrewas, Staffordshire, UK, a location mapped by me and other local contributors for the benefit of all."
http://www.openstreetmap.org/#map=17/52.72818/-1.72908

Interview

Let's start by talking about who is Andy Robinson? What were you doing before you found OpenStreetMap?

Andy Robinson was a civil engineer, living at home, wife and two kids, other than work not really doing much but was supporting my kids with swimming at the time. And they were doing a lot of swimming and going to galas—swimming galas around the West Midlands. That's the reason I got interested and spotted OpenStreetMap because I was looking for a map of swimming pool locations. But I've always had an interest in maps. My father was a geography teacher so he taught me to use maps and I had always been collecting a few—not in any major quantities, just if I ever went off anywhere, I would always have a map with me.

How did you get started?

Well, I had a GPS from walking. And I'd had a go at geocaching very, very briefly, and didn't really get on with it particularly. I was using the GPS for work for gathering information when I was taking photographs. And I was also using it for hiking. And so, I had the tools, if you like, for making contributions so I gave it a go.

There was only really one person at the time who was gathering traces. That was a guy called Alex Wilmer who'd been around the project for a little bit longer, almost from the start. I'm not sure exactly when he kicked off, I'd have to check, but he'd been around the project for a little while and he was working at the time just up the road from where I lived. He'd uploaded GPS traces, sort of going from where he worked to home, that I could see in my local area on what was otherwise a clean sheet, white space. And so, I started gathering additional GPS traces and starting to trace those in as nodes and segments which was what we had at the time.

How did you capture the traces?

That was on a Garmin eTrex Legend black and white, that's what I had.

Were you walking? Flying?

No. It was mixed at the start. I was doing some in the car, some on the bike. I probably tried stuff out locally around the block by walking but mostly I was cycling then a fair bit, albeit not as much as I am now, and driving.

Were you doing this exhaustively, or just on trips, or what?

I was doing a bit of both. So exhaustively only to the point that if I decided to go out and do a little bit then I'd go out and take a few traces. I would always make sure the GPS was in the car and take a trace if I was driving anywhere, so that's really I guess how it started.

And then, how did you get more involved?

Mainly through the frustration of not being able to create or make more of it than that. I'd signed up in 2007 or 2005, and then by the following March, we were still dropping things in as segments but we didn't really have a system of tagging. It was all 'class this' and 'class that', very much a programming type response to how to name things. This was at the same time that Etienne was producing his first Osmarender, which was a means to display a map better than just having these segments and dots on a page. He was producing the first incarnations of a renderer for it and at the same time I was looking to "Well, how can we improve what would show on the map by giving better characteristics to the objects?"

And so, that's when I started working on the very first rendition of Map Features which was very much prompted, in terms of starting it, by you Steve, coming up to Sutton Coldfield and we met, me and, I think, also Alex, if I recall. We were there in the pub. You wanted to find out why we'd gotten involved in OpenStreetMap. And so we were talking about that and in the process of that I'd already, a few nights before, come up with some ideas based on the Ordnance Survey attributes that you find on a standard Ordnance Survey Map. I'd looked at what they generally included and listed them down and created a Map Features list based on the sort of things that typically were being put on maps. That was the start and creation of the initial Map Features page on the OSM Wiki. It's obviously expanded beyond recognition now, but that was the beginning of that process.

So what did it look like back then?

It was very much a single-page of "Ways", that's really what I started off. I had railways, waterways, highways, aerial ways, aeroways – anything that ended in the word way that I could think of to categorize group items. And then I listed those down and made some subdivisions to do in the case of roads - the motorway, trunk, primary, et cetera. Fairly limited, scope of things, not too much detail but reasonably well thought out I hoped.

But not perfect because I definitely got certain things wrong. I wish that I had described the physical objects differently from the administrative. So for instance the highways, I very much should've described them in their physical presence in one form and then dealt with the administrative, as to whether it's a designated foot path or whatever, separately but I didn't. I lumped it all into footway—residential—unclassified, and of course that opened up the can of worms and it still is a can of worms today that doesn't properly deal with the differentiation between the physical space and administrative designation, such as what it is known as, what it is used for, what it's restrictions are et cetera.

So it's really about building the first ontologies for OpenStreetMap?

Yes, I remember at the time suggesting it as an ontology and I was rather shut down by those that were only thinking of ontology in terms of the data set of the elements of Nodes, Segments, Ways. Areas were still being talked about, but no Relations at the time. Everybody on the programming side was very much thinking in terms of ontology as the building blocks rather than perhaps the lesser interest of the tags that we might put on the objects. But I certainly think that it very much was an ontology for the tagging system.

Do you want to talk about what the tagging system was or is?

Well, what it was at the time, I don't think it really was anything. I don't think really that anybody'd given it any thought as to the level of complexity it might need to become in order to display everything you possibly want to display. If you take it at a simple level and you only want to show streets on a map then you don't need to know an awful lot. You could even render everything without any tags. You could just say it's a street. And if you limit it to just streets then that's as far as you go. There aren't any issues.

But of course, as soon as you create objects using the same building blocks of points, segments or ways, you create objects that need more description. So the tags are very much our way of describing what you've drawn on the map.

Over time the tags have become a way of describing all the data objects in more detail and to the nth degree. You can have as many tags applying to an object as you like and they can say whatever you like. You can create a tag that says it's an object that's got a name of something. It's this or it's that. There is no limit to what you can add to it provided that everybody has a consensus that it's easily verifiable, so we're trying to avoid adding tags that don't mean anything. Verifying is very important so that anybody else can go along and see that a road is called whatever it is for example. Or a place of interest is what it is, it does what it does. Keep it easy to verify and you can't bring it into question.

So the Tags are about expanding the amount of information that is relevant to the object and of potential value. You'll never know what's of value ultimately. I mean, there are things that we all put on objects which you think at the time, "Oh, is anybody ever going to look at that or want to?" Of course, you just don't know. Someone in the future may well find that very useful but at the time you put it on, maybe you're not so sure. And there are certainly times when I'm editing when there are objects and tags that I could add and I don't because the level of detail I'm doing perhaps on particular day or particular time just means I want a quick fix and put something in fairly quickly. I don't want to spend lots and lots of time adding lots of rather vague data that may be accurate but I'm not so sure how useful it's ever going to be.

How did the project function at the time?

I think it's always functioned very well. I've been involved in quite a lot of volunteer-type things since OpenStreetMap. It's been interesting to see how there's a lot of similarity and comparison between the different projects and communities that I've been involved with. There's a degree of chaos but at the end of the day things get done. Not necessarily the things that everybody wants to get done or in the right order or to the right depth or necessarily to reach a goal but things do get done. Volunteers come forward and produce something or do something, whether it be adding data to the map or writing a bit of code that ultimately ends up on a routing engine, which is the latest useful thing now to be incorporated into the main OpenStreetMap interface, which is great to see at last. All of these little elements make the thing work and make it enjoyable to stick with.

Does it happen in the timescale you want? Often not. Does it happen in the areas you'd like to see developed more? Possibly not. Some functions come along and surprise you perhaps, well certainly in the early days, once a year, these we fairly major changes that made me sit back and think, "Oh, that's an interesting new direction that we're going in" and all for the good of the project.

So I do think at the time, when the project was kicking off, that it ran well because it had those people that wanted to do things to make something happen. People wanted to do their thing. Not everybody got what they wanted because either they didn't shout loud enough or were not interested in pushing their way through. And other things have probably happened in the way they've happened because people have shouted loudest. But I don't think any of it really is to the detriment to the project. I just think the project is where it's at. It's natural organic growth and the way in which the project progresses, it's not really something that you can force.

Do you want to talk about the conference in Manchester? Why we did it? What happened? How it worked?

Well, I think that was your decision, wasn't it, to run one? And I seem to remember that Chris at Manchester University was keen to host. And so, as I recall, between us, those of us that were interested in making it happen, just got on and sorted it out. I think without having the venue available free of charge to us, it would have been a struggle to get it off the ground. That made a big difference.

So there wasn't actually much to organize and sort out. We just needed to set the process of getting people to be interested in saying something, talking, giving some presentations. We needed to deal with the outreach for people to come, which we mainly, I think, did through the Wiki at the time. And we needed to sort out some activities to do in the evening, which I remember we booked one of the Curry Houses on the Curry Mile and went for a curry.

I think we had around 80-odd delegates turn up and we had a great time.

I don't think that it was particularly difficult to do. It was a relatively straightforward process. There were enough people who were interested in going and wanting to do something and it all just came together. But I think it was an important thing to start. I think that together with the mapping parties were probably the two things that kept the momentum going in those early days, certainly in the U.K., to keep interest to the far reaches, because obviously a lot of people didn't either come to the conference or to mapping parties but they were aware, I think, of them happening.

Do you have any memories of the conference?

I don't really have a lot of specific memories. I remember it being one of the first opportunities to meet a lot of people to put faces to names or usernames that you've grown to know around the country and one or two obviously around the world. So that was the biggest thing that I remember. It was meeting people for the first time. People from all walks of life. All with some interest in either maps or interest specifically in OpenStreetMap and the development of an open source community.

Did you go to the Isle of Wight?

No. I missed that one. I remember you setting that up. That was the first mapping party of course. So, you kicking that off made a big difference, but I watched that from afar. I was sitting here in the Midlands, seeing it happen, watching the amount of data being put in the map specifically by—I forgot his name now, down on the Isle of Wight who lived there—

David Groom.

David Groom, yes, who very much kept it going and finished it, certainly to the extent that we had a finished map of a place at that time, making it look complete. He did a fantastic job. I think Isle of Wight and the other mapping parties that went on later on—I think they helped by showing people that you could achieve something whether as an individual or just a few of you in a relatively short space of time.

What was the first mapping party you went to and what happened?

I think Mapchester was the first one that I came up to in Manchester, the one after the Isle of Wight if I remember right. That was the first one where I did the "cake", where I drew a set of concentric circles to represent the city and divided them up into segments so that we could work out where people were going to map. And we used that—I've still got it somewhere, as the means of reducing the amount of duplication in terms of mapping.

There were some things that I found that don't work in doing mapping parties and I think this still holds true today for a lot of areas. I think perhaps, in some locations around the world, they've been a bit more successful. But I found that giving people a GPS device who have never really understood what OpenStreetMap is, have come along for the first time, give them a GPS, tell them to go out and walk the streets and gather a bit of data, doesn't work. It doesn't work because they often aren't really interested in editing the map data. They're interested in the process. They're interested in the product—the end result. They're interested in the idea behind something like OpenStreetMap but they don't really understand or want to get involved in the nitty-gritty of creating it. The only way mapping parties work are when you have people who are fairly dedicated to wanting to both collect and contribute the data, then you can get a lot done. Just handing out GPS's and letting folks walk around and gather a bit of information isn't a lot of use and certainly is not very productive in terms of time.

But I remember it, like all the mapping parties that went on in the first few years, I remember them all fondly. Quite a lot of them were attended by a number of the same people and we had a good laugh on the Saturday night. Usually, it was a full day mapping on Saturday and certainly a half day on Sunday when it was a weekend one. I remember them all really well as enjoyable places to be. The other thing is, of course, you get to visit places you've often never been to before and certainly you get to visit at a level of detail that you've never done before in these places. I'd been to Manchester prior to that, a few times. I certainly hadn't looked around in detail. I can say the same for just about every place I've been and done detailed mapping in.

Do you have any other memories from other mapping parties? I remember Bath, for example.

I remember Bath for the fact that there were some hideous hills to go up and down on the bike. That was the thing I remember about Bath. I remember the Lake District. We did one in Windermere, based in Ambleside if I remember right. I also did some mapping in Windermere with the same issues as Bath, bikes, steep hills and we had to go up and then back down them. If it was a cul-de-sac, that was a killer. I remember some places where it poured with rain and so it wasn't exactly very enjoyable at times. They were all different in some way or the other but there was an element of fun every single time. I wouldn't have gone along and been involved with them or organized them unless I was enjoying doing them.

Were there any similarities amongst the people that were going?

I think there are one or two characteristics of those people that really get deeply involved and it's often, I think, because that's the main focus and interest in their life at the time. I could say that probably the same for myself is I didn't have any other major interests other than this as a hobby. I did other things but this was a major hobby and therefore consumed a large amount of my spare time to do it. I think that the people that turned up to mapping parties were of a similar mind. They didn't have other lives taking up a large amount of their time. So they had a good chunk of time available to dedicate to OpenStreetMap, to be able to either gather data or edit, because it can be a very time consuming process. If you want to make a big change to your local area or in the case of mapping parties, spend a couple of days mapping, the amount of time needed to edit that data is quite a significant outlay of time. So I think all of those people who came along and were editing and therefore coming along quite regularly to mapping parties were those people who were dedicated to the project and had very little else going on in their lives to distract them from that process.

My memory is a lot of cyclists, as an example. Maybe I'm misremembering.

I certainly upped my cycling once I started contributing to OpenStreetMap but I did it because it was the fastest way to gather data. It was much quicker to cycle around than going in a car especially since it's very impractical to gather extra data when in a car. You can drive the streets but you can't do much else. On a bicycle, you can take photographs even when you're cycling. I certainly managed to do a bit of that, and even if you do have to stop, you can just pull to the curb, take a photograph and carry on. So there's very little delay.

Walking is great. Fine for doing micro-mapping stuff. Nowadays, if we're taking house numbers and doing very detailed stuff then being on foot has brought a lot of benefits, although I think you can still do most of it on a bicycle. But for the early days where we were just trying to get coverage, certainly in the urban sprawl, a bike was much better. Outside of the urban sprawl, arguably a car is more efficient for getting the long country lanes and what have you. I don't know whether or not cyclists were driven to OpenStreetMap or whether OpenStreetMap drove people to more cycling. In my case, I think it's probably the latter.

So how do we get from here to things like Mappa Mercia?

Well, Mappa Mercia really happened as a result of just socializing and the need to have a local presence that people in the local area could identify with. So in the case of the West Midlands here, it was very much talking to the local authorities, talking to one or two other bodies. If we were just talking about being volunteers for OpenStreetMap, it didn't come across very well. So we set Mappa Mercia up as really just a means to have an identity in the Midlands that we could use as a promotion tool for OpenStreetMap locally but also as a showcase for local things.

So we created a gritting map for the winter and that's still out there. We still keep growing that periodically and do a bit more on it each year. It was picked up by one or two of the local authorities who were using it because they didn't have an online map of their own that they could use to show gritting routes where roads and streets which are covered in salt through the winter maintenance period to keep them clear for traffic. So it was that and grit bins – the location where your local grit bins were. So it was something that wasn't on a map at the time. Nowadays, most of the local authorities have a map of some sort with these things on but back when we started it didn't. So as I said, it was very much a means of communicating what was happening locally to the local audience rather than it being either a national/global project.

I'm curious, I understand why you'd want to know with gritting routes but why would I want to know where the bins are?

So that you can go and steal the salt.

Are you serious or joking?

I'm partially joking because what happens in reality is these grit bins get emptied by people for shoving salt on their own driveways in a lot of cases if you live near one. The reason for putting them on was because they're there. So, therefore if you put them in OpenStreetMap, they're there. They just weren't on any other map. It's a bit similar to postboxes where all these post boxes and telephone boxes and whatever get put into OpenStreetMap. Unless there's a map that shows them, would anybody really take care or have an interest in them? Well, it turned out actually that gritting routes and grit bins were quite interesting for a short time because we got ourselves onto the local BBC news when we put it up. So clearly, there was some local interest in having that and for local people to know where the routes were and where the grit bins were so that they could know what routes were gritted or not.

What was interesting, is that it appears grit bins have been placed where people have shouted for them the most. I guess, it's the same with post boxes—the need for post boxes is where people had a need, or pushed for them in the first place. Interestingly, in the new housing estates and what have you, you tend to find very few of either post boxes or grit bins. The density is much less. I also noted that those roads with a bit of an incline where a local councilor either lived or previously lived seemed to have grit bins. So, again, the argument that if you shout and say that you need something and you've got a bigger voice then the others, chances are your local authority or Royal Mail or whatever will listen to you over time and take notice and maybe do something about it.

So it's really about the sort of democratization of place, I guess?

Yeah. I mean, in the back of my mind when I was doing and thinking about grit bins, that idea really wasn't there. It was just a fact that we got all this data going in because we were just gathering it as we walked or cycled around, and the realization that there wasn't a map out there at the time of gritting routes and grit bins. And so, I wasn't really thinking "Wow, we need to get this out." Unlike the post boxes which had been a bit of a drive to get post boxes linked up with the Royal Mail's database with our own that we could see where they all were and that the references for them were right and all the rest of it. There was a bit more of a drive on that. But grit bins locally here weren't an issue, nobody was complaining about them. So it was just a fact that it was an opportunity to put them on the map at the same time that we did the gritting routes. The gritting routes were the interest, I think, locally.

That's pretty funny. Allright, so do you want to talk about the OpenStreetMap Foundation?

Yes, why did we need it in the first place? Well, it's exactly the same argument for Mappa Mercia starting really. You could argue the Foundation was only needed as a vehicle for, in the case of an organization like OpenStreetMap, being able to handle a bit of money coming in - be a body that other organizations can both communicate with and be responsible to.

Until the Foundation was set up, there was always going to be this difficulty of "Well, who should I be communicating with?" I mean, obviously, in the early days, it was you, Steve. So that was fairly straightforward but as it got bigger where did that communication need to go? So I think it was a logical thing to do to set the foundation up just as a vehicle to allow that to happen.

But of course, once you set it up, everybody thought "Well, what is it going to do? What is its role?" So remember we had the initial IRC meeting to sort of define what it would do and get things moving. And we set the aims, which I can't repeat off the top of my head but they were there—aims for an awful long time. I think they've changed a bit recently but the aims of the Foundation being something to help promote the project and not interfere with it really and act as its backer, if you like. I still see that's what its main role should be. But I think it could do an awful lot more if there was enough will to make it happen at the Foundation level.

But the Foundation membership is too small, it's not a big enough organization in its own right to really make a big change to what OpenStreetMap is and does. If all the members or the vast majority of the members of OpenStreetMap Project became OpenStreetMap Foundation members then I think we'd see a lot more done for all sorts of things. But that's not the way that an awful lot of people in the project appear to see the role of the Foundation. I think a lot of them feel they'd rather not have the Foundation at all except that it just deals with some of the administrative stuff of providing servers that make the project run et cetera.

Has that changed over time? What were your memories of the early Foundation?

Well, I think the early Foundation had noble aims to just provide a facility - just to make it work. And certainly, I remember the early meetings and board meetings were very much on feeling our way about how we could help the project. It wasn't really to try and take over the world. It was just to try and make it a better place, to allow OpenStreetMap Project to do more, and that was really the start of it all.

I don't really think that's really much changed when I look at what's happening today, in the board. There are the same issues that were being discussed all those years ago. I get the impression that things haven't changed dramatically. They're still the same sorts of issues that are being discussed. There's a little bit more to deal with in certain areas. Certainly there's more correspondence with respect to things like copyright and licensing than there were in those early days. And obviously, there's all the backup that goes with dealing with legal people, dealing with accountants as the monies come in, dealing with everything else. So just the day-to-day running takes up time. I think that the same issues are there today as they've always been.

What was your period on the Board and what did you do?

I can't remember the exact dates now. I was there three years. I did two years as secretary and one year as treasurer before I hung up my boots.

What were your duties?

Secretary, very much a company secretary, so dealing with the administrative side of the company, making sure that our board meetings ran and the minutes for them were sorted, that sort of stuff. Making sure of the filings with Companies House were done properly, that sort of stuff. So, the administrative side very much for that. Then the primary thing for the treasury chair, then and probably still today, is dealing with the annual conference as more goes on to deal with the money side of the conference than anything else that happens during the year. So, most of the rest of it is dealing with a little bit related to the service and any approvals for money to spend on infrastructure.

Okay, so what was the most surprising thing you've seen in OpenStreetMap over the years?

As I said earlier on, every so often there's been something that's come along that has made me realize, "Well, this is going somewhere." I mean, you used to say a lot, "We actually might be doing something here." Well, there have been a number of things that have made that happen. And whether it be, if we go right back to early days, things like Osmarender was a step forward in being able to visualize what you saw on the map. Okay, things have improved with Mapnik. That being integrated into the project made a huge difference. Just the people taking on the work of the cartographic side. Steve Chilton's early work of getting that cartography for Mapnik right made a huge difference too because these are all things that made it usable. I think the ability to go from purely segments to ways. They're all building blocks that have allowed things to happen. They allowed us as mapping people to get across what we're doing, which is just dragging bits of information out of the wilderness and sticking them on a page for somebody else to make sense of.

I don't think we do it perfectly because we haven't gotten to the point where everything is in 3D yet. We haven't gotten to the point where everything is accurately mapped. And we certainly haven't got the level of detail that's uniform across the planet. So there's a vast amount still to do and I always say to everybody that "the more we put in, the more there is to do," because when you get down to the micro level, if you try and do it all at that level, there's just a vast amount of data still to add.

So there are incremental changes that I've noted over the years that have made a big difference to OSM and have kept me interested. I don't think many of them are groundbreaking, although I thought of them as groundbreaking at the time because I hadn't thought of them myself. The reason I thought that is that you always think "Well, why didn't I...?" Not that I could have delivered, but why didn't I think of that as a way to suggest to people to move forward because that would've been a great idea to have had even sooner than we did get it. So there were lots of little moments like that, especially on the editor side of things with JOSM. It meant getting work on JOSM underway and we've been told that its still used today.

And when I compare it with some of the CAD stuff, I still think that there is a lot of difference between what we achieved with JOSM and some of the CAD operating software. The industry is moving forward, away from traditional CAD, into BIM and 3D modeling and all the rest of it. I think that ultimately that's somewhere we'll have to go with OpenStreetMap. I think we will have to be considering how we model in 3D eventually, perhaps not yet.

Or maybe someone will to come up with the killer editor that makes it easy, a bit like a SketchUp for OpenStreetMap, that allows 3D editing to happen easily and quickly but it will be embraced by the mappers who spend a large amount of their time currently editing with whatever tools they use. They will see that they can switch and do even more with a better tool without having to learn the whole thing from scratch. I think it will take that level of input to make such a big step. But I think that's ultimately where it will end up having to go because that's the way the rest of the world is going. Its not specifically creating maps but is creating objects in space which is really what we're talking about.

Did you see that today the Ordnance Survey announced that they're releasing a large open map?

Yes, I've seen that and it's interesting. I remember when you started out in all of this and we talked a number of times about the pressure that OpenStreetMap brings to bear on organizations like the Ordnance Survey.

I think the open data community has not been very good at achieving what by action has been achieved by OpenStreetMap so it's not surprising they would move in that direction with time. Their initial open data release a few years back came as a bit of a shock to me. Regarding the latest data set it will be very interesting to see what it means for OpenStreetMap in terms of whether we want to use much of it or whether it's a benefit.

Certainly, as a tool, we have found most of the products useful. We've found the data itself a bit hit-and-miss at times in terms of its reliability because it's not always as accurate as we would like. The informational side—street names and things like that—are not always correct in every case. As you grow up you tend to consider the Ordnance Survey to be a definitive map but in reality it's not.

It's kind of interesting. I mean, it took a decade but I'm not sure how—

It's a small time. A decade is a very short time in the cartographic world. If you think that the Ordnance Survey started 200 years ago and most of the mapping that it had 200 years ago hasn't changed a great deal. They have master map, for which the level of detail obviously is much more, but the positions of roads and boundaries hasn't changed a great deal in 200 years.

There's been a big change and shakeup in this industry, across the board, not just because of OpenStreetMap but because of Google and everybody else that have been very active in aerial imagery as well. The whole scene has changed. Everything has changed within this sector. I knew nothing about the sector at all other than collecting a few maps ten years ago. And now, I use it on a daily basis as part of my work.

Right and it's also interesting that one of the initial goals might have been to open up map data in the U.K., but now that there's government open data in the U.K., it's almost irrelevant. It's now about the entire world.

Yes. We have some huge areas that I've not even attempted to even look at, let alone edit in, these still need an awful lot of love. And there's still huge potential to obtain new information, whether it be aerial imagery, whether it be existing governmental data, to help enhance what we have as a map or what we have as a collection of data all in one place because it's not just about this OpenStreetMap being open, I don't feel. I think it's also about having it all in one place because there's an awful lot of disjointed data out there.

I mean, you only have to look at the local authority here in Birmingham and know that they have little bits of data buried away partly in GIS, partly in manual records. It's spread across the authority and one department doesn't know what the other's got. It's a complete mismatch and it would be lovely to be able to take all that data and select what's relevant and correct and right or good enough and put it into OpenStreetMap so that it's only in one place and then it can be used and mashed up together much more easily. It's impossible to do a lot of that at the moment and I think that it's even worse in other countries where the data's just not open in the first place.

What's your involvement in OpenStreetMap today?

Much, much less. I've, probably in the last six months, dropped away from the project more than I have at any other time. It's not that I've lost heart in the project, it's just I've got some other interests which are vying for my time. I'm cycling even more now. I'm road cycling in a club so that takes up time,

The initial reason for getting more involved in the cycling club was to be able to go out further distances to do more mapping a bit further out; get out in the countryside. I was a bit fed up of doing urban sprawl micro mapping all the time. The irony is that it's a cycling collective, and has a very similar model to OpenStreetMap, being run by volunteers. No money changes hands. Everybody volunteers and does different things. I look after routes because I'm a mapping person. It's grown significantly in the last year, and that's taking up my time and interest as is a house we recently bought up north.

Don't you already live up north?

No. I live in the Midlands. That's not up north. So this little place is up in Cheshire and is another area that needs more mapping. The roads are all there and what have you, but it's an area where I can rejuvenate my interest in doing the bigger picture stuff rather than just the house numbers.

What do you mean by "up north"?

Up north, it's not quite Yorkshire but yeah it's a north part of Cheshire, south of Liverpool.

Okay, so what should we have talked about that we haven't?

Obviously, there's politics but you get that in any project. I don't think it's any different than any other walk of life. There are politics in every organization. It's just that in a project like OpenStreetMap where there isn't really anybody with overall control it really is those that shout the loudest that tend to have some level of sway because if they continue to shout, even if you don't necessarily agree with them, you tend to have to listen.

I think in the early days, I was quite happy to respond to and talk and comment on just about anything that came up on the mailing lists and give my opinion. Nowadays, I very rarely do that because an awful lot is rehashing stuff that's been said many times before in a different way. And so, there really isn't a lot more to say.

If anybody joining wanted to go read back in the archive, they'd probably find a very similar thread from five years ago or whatever. So the things that get talked about I think are the same. The same issues come up and they'll never really ever be resolved because there'll always be those people with really strong opinions in the extremities, one way or the other, who will always shout the loudest. Those in the middle ground are the ones that tend not to say much at all. Occasionally, things will sway one side or the other in the same way that politics sways one side to one party or the other with time.

So I don't think that OpenStreetMap is any different from life and communities in general in that there are those people who shout out louder. There are those people that bring things to the table and are more forceful in getting them achieved. There are some people who get themselves into positions that give them a degree of ability—power if you want to call it in politics. But whether it be somebody administering a server, or someone looking after the style sheets for a map, whatever it is, if you've got that little bit of extra ability because you've got some control, that gives you a little bit of power, that gives you the opportunity to do some things that other people wouldn't get an opportunity to do. So that tends to shape things over time and it also, perhaps, makes it more difficult for others to get involved in some aspects of the project.

Since day one there's been a call out for more developers and more code writers to get involved in the project, but those people have to have an interest in maps or geographical data in order to do so. They're not really likely to get involved unless they can see a really interesting thing that is needed. And it's difficult to communicate those interesting things to people who are not interested in the overall project. So the politics I think will always stay the same. There'll always be people who will be very outspoken. There'll always be people that rub other people the wrong way. You get criticized some of the time for rubbing people the wrong way. Well, that's just the type of person you are, Steve.

Thank you. That's great. Let's definitely include that in the transcript.

That is you. Everybody has their character and without the character–you need somebody at the start of these projects. I'm convinced, having seen it work with other things as well. The cycling collective is a prime example that's close to my heart right this last couple of years as they were last year.

You started OpenStreetMap not only because you had the capability to start it but also because you were passionate about telling everybody about it and making things happen, whether it be you doing it or you saying to somebody else, "Will you do that?" You weren't afraid to push people to do things because that's the nature of your character, and that tends to cause some friction with some people over time. When people rely too much on other people to do things, that can cause issues with over time. I think it's absolutely fantastic for getting things off the ground.

There's one particular guy who's the chairman of the cycling collective who is very similar to you. He is a person that will push people to do things. He'll say, "Are you going to do that? Can you get on with that?" And it's getting those people started to take on a role, doing a little bit, that seems to take a project from just being an egg to turning it into a chick. It doesn't get it to being a chicken, but it certainly gets it to being a chick and beyond that I think that's where it's best left, because the longer the benevolent dictator stays in the position, the more difficult it becomes because there is the danger of too much control.

You tried to hold onto control. You recognized eventually that the danger is if you try and keep the control, you can't do all the things that need to be done. The best way to allow something to grow is to let it go. And my friend in the cycling collective is at that point now where he needs to let it go or he's going to find that it's going to be difficult in the future to allow the thing to change and morph and organically become a much bigger thing.

It's still important to have the drive. You've still got the drive for OpenStreetMap. It's just that you're no longer in a position to be the same sort of advocate that you were when you kicked it off. There'll still be people who'll come and ask you to talk about OpenStreetMap because you know more about it than anybody else as the founder of the thing, so that's a given. But I think there's a benefit in stepping away.

But it does raise a question. It's that, when you get to a certain size, what is the best way of having focus and getting something to progress at the same sort of rate of speed and of enthusiasm it did in the early days? How do you stop the fatigue setting in and keep it very current and active? You can do it in commercial businesses where the dictator at the top or the board can dictate what happens and drive the focus. It seems to be very difficult to achieve that in large open source or community-type projects. I guess that's why Communism failed.

Is there anything else we should have talked about?

There must be stuff, Steve. I'm just trying to think what—I mean, you know, in ten years, not quite ten years even for me but ten years for you. Well, it's nearly 11 years now. But it's still a long time for me. There's an awful lot of stuff that's gone under the bridge.

I think there is one thing that's interesting and that is that as I said a minute ago that I'm not doing as much now. These last few months, I'm not doing as much as I had been in OpenStreetMap. But if I think now about all those people who were influential at the start of the project, in the first say three or four years. Most of them are still involved in some way or another, or certainly around in the scenes. Okay, there are some names that I don't hear about anymore, have moved on and joined different organizations. You can argue that you've moved on into other organizations, and are less prominent within the project itself but there's still an awful lot of people who were around in the early days who are still there today.

I don't know whether that's a good or bad thing. I mean, if they're contributing valuable stuff, it's got to be a good thing. But at the same time, the old guard doesn't perhaps allow some things to happen that might be good for the project. That's an observation and I don't know how you overcome it. You can't force new blood into these things but there are people who have been active and are very vocal and do certain things. And they're the same people who were there seven years ago.

Do you have some examples of some of these things that might be better?

I'm not so sure that they would be better because I don't know what would come if things were different. But all I can think is that if I looked back ten years ago, you and one or two others were very instrumental in getting this thing off the ground. You did the large share of it but there are other people that input in a major way, whether it be a Tom Carden or Ben Gimpert or whatever and Matt Amos. Matt's still involved of course. You've got people like Frederik Ramm still in, been around for an awful long time. Richard Fairhurst, of course, still around doing stuff.

Now, if you take those people away and there was new blood, if you take away someone like Tom Hughes doing all the, you know—what an amazing guy he's been for doing all the administration stuff we've got on the servers. Without Tom, where would we be now? Or would we be any different? Would it look different? Would the project be further forward or backward from where we're at now? I don't know. I can't answer that but it would definitely be different. So the question is, how do you shake things up now and again and allow some new blood in to bring a fresh face on things and do things a little bit differently? Because I think that change is always a good thing. It's wrong to think that you should always just chug along the way you've been going.

We have the graphical stats pages that look the same, month after month. The look of them is identical. It's still an upward growth. It's not quite as steep as it was but it's still an upward growth but the long tail still looks roughly the same. So nothing really changes and I do wonder, "Well, does that mean it's just going to slowly, over a long period of time, wither away and just become a static part of the tail that isn't changing much? It's not lighting up the world. It's not producing a new iPhone 10 or whatever it is. What will it take to make a big difference to the direction of something like OpenStreetMap?

We know that in Germany, the national press around the project made a big difference to the steep increase in user interest. That happened pretty much overnight, a few years ago. That hasn't happened anywhere else. It hasn't happened in the U.S. I'm still amazed. I go to the U.S. periodically and the level of detail in OpenStreetMap is almost as bad as the day it was when the TIGER import was done. And it still astounds me that there hasn't been any take up. I understand partly the reason it is because there's so many other resources out there that people can use that really means that they don't need to spend time editing OpenStreetMap. But it surprises me there aren't more people out there who would do it just for a pastime, be interested in doing it.

Grant Slater

Bio

Mapper since late 2006. Been part of the osm operations team since 2007. South African living in London. Every holiday to South Africa is a multi week mapping expedition. Have over 15 years experience running and managing n-tier linux server infrastructure. Deep/low-level server hardware experience. Leads osm capacity planning and hardware upgrade cycles. Dutch-bike Cyclist (batavus) Day job: Lead (Dev-)Ops engineer for a large London digital agency. Banking, Utility & Healthcare industry.

Favorite Map?

> *"The aerial imagery I put together for South Africa. http://aerial.openstreetmap.org.za/*
> *Was a great learning curve and was reasonably fun to put together the 4TB+ of data used. Otherwise the early topographic series maps of South Africa I find amazing."* http://grant.dev.openstreetmap.org/ZA-Topographic-Out-of-Copyright/original-set/example-slippy/

Interview

Why don't we start off with what you were doing before you found OpenStreetMap?

I think I was just bored in London actually. I wasn't really doing much. I saw the project and saw it as something quite interesting and decided to get involved. I think I immediately paid for an OSMF membership even though that was brand spanking new because that was back late 2006, so it was really new at that stage. Yeah, so I wasn't doing much. I was a bit bored and looking for something to do that could be interesting.

What were your first thoughts?

It looked like an awesome project. I could understand where it came from. I could see its use for more of an outside just the U.K. Being from South Africa, I could see it being useful for South Africa as well. I decided just to give it a go and see what I could map. I think the first thing I mapped was actually a railway line in South Africa that I travelled on just a little while before with a GPS, I remember too.

Grant Slater

What was your first involvement in the project?

The first time I really met people was—I went to the—damn, the mapping party. You weren't at it. It was the small county. What was the county up north?

Rutland.

Rutland County. Yeah, so I popped along to Rutland County. I was so excited I rented a car and drove off not quite sure what was going to happen but yeah it was pretty good. I met a few people that seemed quite interesting. They gave me a patch to map and off I went. I brought a cycle because it seemed like what people used to map. I took a little square very nervously.

What are your memories of it?

It was good fun. All the people seemed interesting. I met this crazy guy called Richard Fairhurst who was missing teeth at the time. Initially, I thought he was some like hobo guy but he was actually pretty cool. Once he started speaking you could quickly tell that he's not some rough guy from a rough part of town.

This was after his cycling accident, I guess?

Yeah. He had had an accident and knocked out his front two teeth. He looked like quite a rough character.

That's funny. What was your involvement after that?

I'm not too sure what was the next thing I went to but it was probably down in Dorking. There was a Dorking mapping event. I met you at the Dorking event and a few other people. It was good fun. I don't think I got all that much mapping done that weekend but it was still interesting to meet some more characters of the project – some more interesting personalities.

Do you want to tell any stories?

I met Nick. I think, Nick Whitelegg – the walking guy. I thought "he's quite an interesting fellow." He loves his walking. I gave him a lift around his walking spots. It was quite an interesting weekend. Again, I got given a square that was way too big for me. I tried as best I could to cover it. Yeah, it was good fun.

And so, how did your involvement in OSM evolve over time?

I'd started looking at some of the architecture things. Because we were running MySQL then, occasionally there were problems like the database used to time out and things like that. I asked you for a—I forget whether or not it was the MySQL configuration file or table dump or something like that and you made one. I looked at it and made a few suggestions.

A few other people got involved around then. Jon Burgess and Tom Hughes had become involved before me. I don't know exactly how I got given some server work but somehow I got given. Probably that's it, Nick Hill. He was eager to hand over to some people and I think you suggested me to him after I talked too loudly about MySQL and a few other things. I went to UCL with him. He showed me around. Very quickly, I was given the keys to everything.

Is that literally the keys?

I was literally actually given the keys. The machines used to be locked with front panel keys. I was given a copy of a set of the keys fairly early on. I was very nervous about being involved—looking after someone's servers.

I knew Tom and I knew Jon were quite involved. They were sort of the main figures and such—Tom doing some of the app code and Jon doing the tile code. What I did was, I said to them, "Well, why don't I come and see you?" and we could discuss plans and what their strategy was. So, I rented a car again and drove off and visited Tom and met him at his house and chatted about where he thought the project was going. What new servers did we need? Did we need to spend more money? I had a good chat with him and then went off to Jon. We only lived like about an hour away. Obviously, the chat was more focused on the tile server and what he thought we needed and where we're going with that, quite interesting. It was more really an introduction to each other. And then from then, I think I started buying stuff pretty soon after – buying new hardware and expanding things.

Where did the money come for that?

The initial money came from the Foundation. We weren't talking much money. I think I put in some RAM here and I put an extra CPU or something there because the hardware project was pretty basic. It was all just repurposed desktop machines.

Your old UCL server was still sitting in the rack at the top of the rack and plugged with some hard drives in. The machines were all just your standard desktop machines, your standard desktop hard drives and ungodly IDE cables all over the place and some interesting things that were being set up. The money came from the Foundation. I just spent a minimal amount of money just buying RAM and scraping around eBay to find some of the cheapest RAM I could and just started slowly building things up.

Do you want to talk about how the server infrastructures evolved over time?

It took us probably at least two years to start, dare I say, getting rid of some of these old desktop machines and starting to replace them with, dare I say, proper server machines. The desktop machines were just painful to maintain because if we wanted to replace a hard drive, which happened quite a lot in the beginning, or upgrade a lot of the hard drives to faster and faster hard drives, we used to have to completely disassemble the machine and plug in the hard drive or check if it would boot and change file settings.

Fairly early on, I think we got donated two HP servers, 1U, pretty good little servers. I think, for one, I had to buy a guy a pint, and the other one was just given to the Foundation. I got those up and running. They were pretty nice. Well, at that stage, pretty nice machines for us. Tom got some of the code running up on them. Slowly, we started to get one or two more of these machines.

Since then, we've standardized completely on HP machines and many of the core things. If it's a service that is replicated, then I'll buy a few machines off of eBay, spec them up and everything runs great on these machines. If it's something important like a database server, we'll go out and buy a brand new server or get that all spec'd up, tested, run it for three months as a secondary service and then move to it to the primary service after that. So, hardware-wise, we don't spend very much money considering how much hardware we now have. Globally, we probably have 30 servers now compared to when I started when we probably had about five or six really low-end machines.

What do all those servers do?

The main website is made up of eight machines now. It's three front-end web servers. Three app servers. And then we've got two database servers behind there running a replicated pair. So, that's the website. And then we have a file storage machine, so that makes nine. Then, we have two rendering servers. And I think about 14 caching servers in front of that with intercache sharing.

So then, a West Coast U.S. machine will access an East Coast server and it'll access a server probably in London that will do the rendering if the tile hasn't been rendered. After that then, we have a whole mixture of machines, some doing help, some doing the OSMF Foundation blog, which used to be the OpenGeodata blog. So, we have a lot of machines. We have a list of them and what their tasks are.

What have some of the challenges been over the years?

We do things quite slowly. We build the servers quite slowly. Probably the main challenge is getting the software developed and finding out how to do things. Well, the best part for getting certain things done like–I'm not too sure actually.

Well, I mean, have you had any fires? I mean, have servers gone down?

No. So the biggest challenges we've had is we have very generous server hostings from two universities in London. UCL who initially hosted the project and then we've now got server hosting from Imperial College London. UCL is very generous but we occasionally have power problems in the data center that we are now in.

They've been doing power maintenance on the machines. The services that are run there, before the power goes out, we have to move them to other machines. We do that all via automated systems or deployment systems that we have but it's been quite a struggle when we have a full data center outage for a weekend or so. But we've been handling it reasonably well. For example this last weekend, we had an outage at UCL for the whole weekend.

Why is this still going on? This is like five years later?

I know, it's a bit crazy. We really do appreciate the hosting that they give us but we don't run any primary services there now. Everything that is at UCL is replicated somewhere else or at the flick of a switch, it is replicated somewhere else. We keep getting promised better deals from UCL. They want to move us to a new data center but there's problems involved with that in that there's not enough rackspace for us.

I'm curious, is it still the same guy managing it?

Kind of. We moved from—I forget what the old building is but we've moved to—

Wates House?

Yeah, we've moved from Wates House to Wolfson House. And so there's a new group of people at Wolfson House that are helping us there but they really don't like us much because some of the older machines are quite power hungry. We've exceeded the capacity in the rack and then the UPS keeps complaining that we're using so much power and then we have to—So, over the last six months we've been upgrading some of the systems to a lot more power-efficient machines.

How has the demand on the server infrastructure changed over the years?

It used to be that we would be quite nervous when there was a posting on Slashdot, or a TV report or things like that. We used to have major traffic spikes that we absolutely couldn't handle. We had very little spare capacity. Now, we don't worry so much about it because we've got enough capacity spare in all the machines and CPU and RAM and all things like that that we can handle massive traffic spikes. We've been on fairly big news sources over the last few years.

We can watch the traffic double or triple even but the latency only goes up a tiny little bit so we can handle it fairly well. Some of the servers just struggle more than others but the primary website has no problems at all. Getting that right probably has been one of the trickier challenges. Getting it so that we do scale better to handle a load like that and getting all the services that people access and use to be scalable.

Can you talk about that?

Well, so we've moved. Some of the bottlenecks traditionally have been the database which is slow. What we've done is we've replicated there and then we've moved some services out to a memcache layer so log-ins and all of that are in a memcache layer.

And then different pieces of software have had their own issues. The Nominatim service used to really struggle because every lookup would be a straight request to the database. What has become a problem over the last few years is we have people that make mobile phone apps for tracking vehicles or whatever, they abuse the system. They've set their app to display what suburb the mobile phone is at the moment—the tracking software. It does a Nominatim query every ten seconds, or three seconds, or one second.

We've had things in place that limit abusers like this. We've done it for the tile servers. We've done it for Nominatim. What we do is we initially just slow people down if they're flagrantly abusing the service. And then eventually, at a point, if they continue to do that then that IP address or that client would just be blocked after a while. And then within a few hours, an automated system will unblock it again. So yeah, that's been fairly tricky to get right and find the right levels for some of those things.

People reading this interview might think that you could just go to Amazon and sort of there's a dial and it goes up to 11, right? And if I want to scale, all I need to do is turn up that dial. Do you want to talk about how right or wrong that might be?

We've taken a decision, as part of the Operations Group, that we want the server infrastructure for the project right or wrongly to be something that we control and which—that's the wrong way of saying it. We don't want to be

reliant on anyone else for continually running the services. We don't want to be getting a free service that maybe runs out in six months' time and suddenly we can't support OpenStreetMap anymore because the free service has run out. All the core services that we run like the website and the API and some lesser services like the tile rendering all run on OpenStreetMap-funded hardware, all OpenStreetMap-donated hardware.

We have to watch the capacity on those because if you have a mobile phone that decides to download tens of gigabytes of data, that's going to cause significant problems for other people potentially wanting to view a map or use the data in other ways. Obviously, you can always download OpenStreetMap's export data called the planet file and then you can do anything with it you want. But when using the API services like Nominatim or downloading thousands of tiles, it puts a strain on the servers. So, yeah, we restrict some of those things because the capacity that we have is limited and is shared between multiple people.

Aside also from the non-technical decision to choose your infrastructure yourself, right?

Uh-huh.

How about also on the technical side. Let's pretend that you moved to Amazon AWS, would there literally be a dial and you could just do that and everything will get better or not?

We looked at Amazon web services a number of times and a lot of the software architecture would have to be re-designed and re-done. Because we're all volunteers, that is actually a more expensive resource than actually buying hardware. Having developers write the software to be able to run on these sorts of services. I've said it a few times, if there's something that we're short of and it's quicker just about hardware or throw resources to like that, we'll buy the hardware. We'll buy it as need but we hold back a little obviously.

Whereas things like running the rendering server setup, it wouldn't run well on Amazon web services because it doesn't scale out as well as we'd like it to. It runs well on a small cluster of machines but we haven't got it to. You'd need fairly big machines and the Amazon web service needs to be able to run a rendering service and we haven't got it to the stage where you can segment the rendering service into "well, this is only West Coast U.S. or this is only East Coast U.S." So you run a full rendering server stack that potentially can serve the whole world. You have to run that service for it to work reasonably well. We can't segment the service that runs on smaller machines across a bigger infrastructure like that. I'm waffling a bit.

No, that's fine. Do you also want to talk about how the architecture of OpenStreetMap is fairly separable? I don't know if separable is the right word—independent. Like, it's not all just in one machine It's also that the tiles are really sort of loose. Yes, that's it. Everything's sort of loosely coupled.

Yeah, so the OpenStreetMap services are loosely coupled. We can detach many of the parts from the whole. Take for example, the rendering service or the Nominatim service, they feed off the planet diff files which are minimally differential files that we export from the database. Those are available to anyone to pull onto to their app. Now the tile service, for example, pulls those and every minute adds it to its rendering database and absorbs this. We can move the rendering services anywhere we like or we can add additional rendering servers as at will. As with people like Mapbox, they take those diff files and they render using their framework.

It's a completely detachable service as long as the core produces those diff files which we consider a vital service which we put a lot of time into making sure that it continues to run reliably. It allows others to take off some of these services and run them themselves if they wanted to. Or we could, in the future, maybe decide we don't want to run our tile server, it was too time consuming, other people can pick up those services and run them.

What's been the most challenging thing over the years of running all of this?

I think the time is probably the most challenging. We've gone through a few phases where we've rebuilt things. We used to admin every server individually and log in and you type all of your commands to set up the software and you'd sort of have a playbook of run this and then run that. Eventually, you'd have the machine running in a day or two with all the software configured and hopefully running correctly.

We went through a phase where we encouraged, a few years ago, back in—I don't know when State of the Map was in Spain, in Girona. We encouraged quite heavily then to start using a configuration management system. There were a few available. Puppet and Chef were discussed. We decided to go with Chef.

Tom Hughes looked at Puppet and looked at Chef. He saw that Chef was probably easier to get up and running. We started running with that and started quite simple. We just moved off one service at a time, add it to Chef. When we set up a new machine with that service on, we did it all in Chef. No manual configuration. We slowly built that up over the last few years.

The Wiki server, for example, is all just click to deploy. You run one script and you have a wiki server installed. Or if you want to install a Nominatim server, we just literally set the machine to have a Nominatim role. I think it takes about two hours for it to download and install the software but it's all completely automated. The machine, within two hours, is a Nominatim. We've used that to move certain things. As we've added services like tag info—so OpenStreetMap now runs a tag info service – taginfo.openstreetmap.org.

What is tag info?

Tag info takes in the planet file, weekly exports, and it analyzes all the tags that people use to annotate geographic information. "Is there a park here or is there a pub here?" It takes all the tags that people use, so the key value pairs that they use and it analyzes them and finds out what are the most common and popular ones. It then cross-references that data against the wiki that we have where people document these tags that they're using and what they're used for, and images and examples. Tag info also then compares the ID editor's database of tags, the JOSM tag database. It compares all of these together and produces a very good summary of what are the most popular tags, how people use them, geographically where they're used. And then tag info's database is also absorbed into ID's editor to suggest tags when people are adding stuff. So, they can go "well, I'm looking for a road" and we give them a few options of the types of road. The same with service—all sorts of services and it just queries the tag info service.

What's been the most fun thing to do?

Just watching the project grow, because it has grown drastically. We're now close to 2 million users. When I joined, I was user 5,000 or something which was pretty high, but we've grown to 2 million. It's been fun getting the project to scale, adding in all these services that are needed to run the project and seeing where we don't have enough speed and scaling up the machines or scaling up the architecture so that we have more machines doing the same role. It's been pretty fun working out solutions that will work to keep the project running.

What is the Operations Working Group and how are the responsibilities divided?

The Operations Working Group is some of us people who have day-to-day access to all the service and write the Chef code and maintain - look at all the graphs and projections of how the machines are doing and where they're going, down to disk space. That is the part of the Operations Working Group. We also then have a few people that are more deep thinking, dare I say.

We have a guy called Andy Allan and a few people that have been involved with running of the services or less involved today but still give really good advice. Together, probably once every three months, we meet up and/or we have a video chat. We discuss, "Is this under resourced? Do we need to buy new hardware here? Or can we run using the existing hardware?" We work out budgets of how much we'll need to spend and when we need to spend it. All those sorts of things. Are we going to have to get another data center to do some of these things? We just try to plan three months, six months, and twelve months ahead.

Is it a committee? Is there a formal division of power? How does it work internally?

There's no formal organization structure as such. Tom Hughes does a lot of the Rails code. He's sort of turned into the Linus Torvalds of the OpenStreetMap Project. I hope he doesn't find that offensive but he controls all the code submissions and he suggests on ways of implementing certain things. And then we have Matt. Matt who gets stuck in the hardware every now and again and helps me out. I'm normally the guy who lifts and carries the hardware, or specs up new hardware with Matt. We get it installed in UCL or Imperial or in some random data center all over the world. We've got, I think three servers in the U.S. and the Netherlands and Azerbaijan and all over the place. We've had servers all over the world primarily for tile caching. But most of the core infrastructure is still in the U.K., unfortunately. We'd like to expand more of that out. We've got some servers in Oregon now and we want to scale that up over the next few months, possibly a few years.

It's really quite remarkable how things have scaled, isn't it?

Yeah, it's fun to watch.

Especially with sort of minimum investment and volunteer time.

The amount of money that we spent is miniscule for running a project this size. Just bandwidth-clustered AWS would bankrupt us within probably a month. But yeah, we keep tagging along and keep things growing which is great.

Are there any fundamental changes on the horizon? I'm thinking and correct me if I'm wrong but the solution has always been "when the database gets too big, buy a bigger database server." Is there anything that's going to fundamentally change?

We've been growing quite nicely with the growth of hardware. Hard drives are becoming cheaper. RAM is becoming cheaper. CPUs are becoming cheaper.

Traditionally, we've had our problem with the write capacity of our database servers because we have so many people reading and writing data back to the main database. We have a limit on how many writes per second and that has been quite a pinch point for us so we spent quite a bit of money buying expensive hard drives and putting them in expensive RAID configurations where we have lots and lots of spinning disk heads and that has been a pinch point. But we're starting to get better at that. We've split the write load from the read load and we can scale out the read load as much as we like. Also, software projects like PostgreSQL which we use. We use PostgreSQL as the database server and have been getting better and better at replication. There are new releases coming out to do mutli-master replication that would, for example, let us scale out the write capacity over multiple servers which is a great feature that we're wanting.

I don't see any major hurdles on the horizon. There are a lot of people that have come and mentioned lovely NoSQL scaling databases that you just plug in a new server and it internally scales out all the hardware. A lot have invested developer time in the set up that we have. PostgreSQL comes and goes out of favor but it's growing in favor as a truly reliable database that big corporations are willing to invest in. We're pretty happy with it. We've got some good in-house knowledge of how it runs, how to optimize it and how to keep it running in a reliable fashion. I don't think anywhere in the even distant future, we'll move away from PostgreSQL.

What were the reasons for moving to Postgres from MySQL?

Back in 2008, we were using MySQL. As I mentioned, we have high reads from the database and high writes to the database when people are updating the map. The high concurrency was starting to become a breaking point in MySQL. It was just locking the database way too often.

What is concurrency?

Multiple people doing things at the same time. We regularly now have mapping parties where over 100 people are editing the map simultaneously and this is happening all over the world. So, having a shared database like this when many, many people are editing at the same time, it just starts to become a problem with concurrency when you are having all these people reading - some people reading, some people writing, maybe editing fairly close data to each other, it starts to become a problem. MySQL really doesn't scale well with high concurrency. It would lock an entire table with data. Whereas with PostgreSQL, it only locks minimal records and allows more people to work simultaneously.

Can you give people a sense of—pick any metrics you want, transactions per second, number of users per day, what those numbers are today and how they've changed over time?

Without looking at some of the graphs, I would struggle a bit.

You can look as well. I mean, like and how many tiles are downloaded a day? All those numbers.

Okay. I'm just going to pull some up quick. Superb. I know we're serving multiple terabytes of tile data. I think it used to be about—was it one of the 12 caching servers was doing over a terabyte of data a month. Wow! So that works out—sorry, one of the tile servers was doing 30 terabytes a month which was the capacity that the ISP offered and then we've got 12 of those servers. So yeah, it increasingly becomes a problem. We had to scale that load up and we have done. Over the last year, we've gone from about 180 megabits per second for the tile traffic to close to 350 megabits per second. So we—

Nearly double.

Yeah, it's more than doubled so it's about 2-1/2 times we've doubled the amount of traffic that we serve, just for the tile service. Now, that's replicated across all the services so each year we're nearly doubling the traffic. With just quite growth problem to have, it's fantastic. We have the project growing at such a rate. I'm just trying to bring in some more graphs. I should've prepared more of these. Or probably, we can slide in a few numbers. We have at least 3,000 people editing every day. I don't have my one IP-wide list and I can't access the one stats—internal stats, a thing we have. Yeah, I have to look at them. Damn it. Sorry, you were saying, Steve?

What, if any, was your involvement in the license change?

I was part of the Licensing Working Group where we spent, I think, close to two years working, asking the question "What the license change should be? How we should transition to it? How we would get agreement from—I think at that stage it was 1 million users. I can't recall exactly but it was quite a project to work at "How do we contact all these people? How do we get engaged with them to reconsider licensing their data?" under, in my opinion, effectively a very similar license to the license they'd already agreed with. So yeah—

Why was it set up in the first place?

Well, right from back in 2007, when I really started to get involved with the project more extensively, there was the discussion that CC BY-SA was too limiting on the project. It had too many ambiguities about how the data can be used. If it could be with print, would it be used with print? The licensing there was a concern. Well, we got the ODBL drafted as part of the licensing working group. We worked with people that drafted it and it was considered a much better license. Yeah, I can't really think more than that, no.

Do you want to talk about the process of the license change and things that you worked on?

No, it was a painful process which I kind of blanked from my mind.

Most of the people reading this aren't going to have any memory of this change. They weren't involved. They weren't there at the time. So you give a historical perspective of why we did this, what happened and what was involved. It's not about complaining about people with public domain or anything. People today who read this book won't have any knowledge of this. This is all ancient history.

Yeah, so the license was a problem with the project right from the beginning. A lot of people thought that the ODBL was too restricting and it was too unclear in how the data could be used. The Licensing Working Group was tasked with coming up with a better license for the project that made using the data simpler and easier for a lot of people to use and particularly in cases where derived data could be used from OpenStreetMap and how you could include other data sets.

We spent a lot of time going through revisions of the ODBL and working out how the license would be implemented, and right down to how we would answer the question, what terms would be—sort of the questions to people that we would send out, of how to get them to agree to the new license. What key points in the transition would there be? Just to get the ball rolling so. For example, we had— when new users signed up after a particular date, they agreed to the terms of the license along with the old one. And then we started asking people that had been part of the project for a long time, whether or not they would re-license their data.

We really got the whole community involved and trying to find some of the people who had contributed data and were no longer involved. We were trying to track these people down and asked them, "Do you mind whether or not we re-license your data?" And also, engaging with some of the community. There were a few very vocal voices that the license change was a bad idea. There were a few personalities who we really ruffled their feathers. But, as a whole, it was by far a minority in the project. There were only a few individuals but they created a lot of noise and a lot of disruption.

Trying to answer the recurring questions from some of these people, we would answer them and they would just re-phrase their question and we'd have to go back and say "No, we've covered this point. Or, we have addressed it here or we've answered it here." It was just really painful.

So, we got the whole community. We had a vote halfway through asking the community "Are we going in the right direction?" and we got the positive backing that we needed. The license change went ahead. I think, in the end, we had contributors of the data. So the key value was how much of the data would the project retain after we'd asked everyone and we'd got answers from it. Data-wise, we had I think 96% of the data could be re-licensed and was re-licensed, which was phenomenal. The amount of data loss was absolutely minimal. There were a few places where we had a few contributors that had done major mapping.

For example, Australia, we had a number of individuals there that didn't agree to the re-licensing of their data based on their principles. I think it took six months for Australia to be back to the same level that it was before the re-licensing. But very quickly, Australia was re-mapped. In Australia, I think, we only carried about 70% of the data through to be re-licensed. But very quickly, we had the country back to the level. And we had people from all over. For example, the coastline of Australia was mapped by, I think, primarily three people early in the project. Two of those people decided not to re-license their data. We had the coastline of Australia re-mapped in a weekend. And then revised and improved over the next few weeks. So it was surprising how quickly the community responded to fixing up the map data.

Did we lose any data in London? And did we do anything to fix that?

Yeah. So we did lose some mapping data in London. One of the mappers, in Southwest London, who had mapped a huge area in Southwest London on his own, traveling on his bicycle I recall. He decided not to re-license his data. I think about 10 or 12 of us went to his hometown and then we started mapping it again. And within a weekend, we had the town very well mapped. Well, in my biased opinion, very well-mapped again. And we had a few events in the area just to improve the coverage and get more data there that we had unfortunately lost. We had some data loss in London but it wasn't a major setback for the project. The project came together. We re-mapped it in a very short period of time.

What were some of the technical challenges throughout all of this?

One of the major technical challenges was the layering of data. I may put in a road and then someone else may improve the alignment of that road or put in more points to make a nice curve if the road curves. And so, one bit of data, for example one road, may be edited by multiple different people. And if one of those people have decided not to re-license their data, it was quite painful. We had to work out "Could we re-license this road or this point or this name?"

Matt wrote a very nice series of tests basically, what we would do in particular cases. If the points had just been moved by someone who decided not to re-license, we would move the point back to its original position. He built up hundreds of these sorts of tests and we turned that into a software package. It would ingest all the data and all the history of the data and work out what we'd do with it and re-build the data with all the people that had disagreed to the re-licensing, not taking their contributions in mind. The software took quite a while to write but Matt is a very talented programmer and we've got some interesting software there.

We had Andy Allan get involved in writing some of the API stuff. Some of that software was written by him. When we eventually ran it live on the server, we'd thoroughly tested the software and seen how it would work. We'd used it only on one region first. Then we expanded what regions and we broke up the whole map into multiple cells and we ran against these cells to sort of remove the contributions of people that had decided not to re-license their data.

How did you feel after it was completed?

It was quite nerve racking. Although we had worked out how much of the data would be retained. We had a very good idea what sort of impact it would. We really didn't know how the community would respond. Some people have invested years of their life making their neighborhoods or their towns or cities look brilliant and have all the data that they could possibly add to the map.

We didn't really know how some of these people would—would they just throw up their hands in the air and walk away and go, "It's all ruined. It's never going to be fixed." Or would they get involved in fixing it and improving it again and making just a generally better map. Surprising, the community sort of pulled together and we got things fixed and got things running again well. In biased opinion, we have the best map data for many parts of the world, thanks to the community.

Do you want to talk about any memories from conferences or mapping parties that you have?

Unfortunately, we don't seem to have as many mapping parties in the U.K. as we used to. I met many people through the mapping parties and we're still quite a close-knit group. Mapping parties are a great way to interact with people. It's good and all sitting behind the screen and editing a map and seeing the world but OpenStreetMap's also a social project and you get to meet other people that share some interests with you but may have many diverse interests as well.

We have people that are interested in history of buildings, or maybe interested in how government regulates roads, or cycle infrastructure. We have biologists and— such a multiple—so many fields and it's great to meet these sorts of people in a public setting and talk about shared interests and talk about how we map things. And, "What are their side projects?" because OpenStreetMap is a side project for many people.

It was really fascinating to meet all these characters and see what they do and see what their passion is within OpenStreetMap. Why do they contribute? How do they contribute? What methods do they use? What they've discovered? What's hidden in their suburb that I haven't yet found in my suburb. All these things, it's really good fun.

What should we have talked about that we haven't talked about?

The one thing that fascinates me of late is how the project is growing in parts of the world where I never thought the project would grow. I'm originally from South Africa and we have a small country in the middle of South Africa called Lesotho. It's relatively a poor country with not much development but OpenStreetMap is now thriving in Lesotho. We have people from Ireland are flying out there every now and again and helping the community map their streets and their buildings and adding in the whole of Lesotho as a project.

So, while we are all dispersed all over the world, we have a common goal and we're building a map for all of us. It's amazing to see how the community does this and how we all are excited about making this map data that we can all use. It's a great resource for everyone.

We have humanitarian projects in disaster-hit regions where they're focusing on making a map better and more useful both for their self-interest and for the community there because the map data are open and we can call use it. They're helping the community leapfrog and use the map data for interesting things. We have the map that's being used in Kenya by people making independent censuses of places that haven't been in censuses before. We have disaster relief projects all over the world where people are using our data and using it for navigation, or rushing to rescue people, or analysis or all these sorts of things that map data can be used for - people are using OpenStreetMap data now for. It's just fantastic to be part of a project where we have made this asset where so many people may benefit from.

Made in the USA
San Bernardino, CA
06 January 2016